Boutique Hotels

精品酒店空间设计系列

Guestroom & Bathroom

客房与卫浴

谢昕宜编　李婵译

辽宁科学技术出版社

Alila Jakarta

艾里拉酒店

Jakarta, Indonesia
印度尼西亚 雅加达

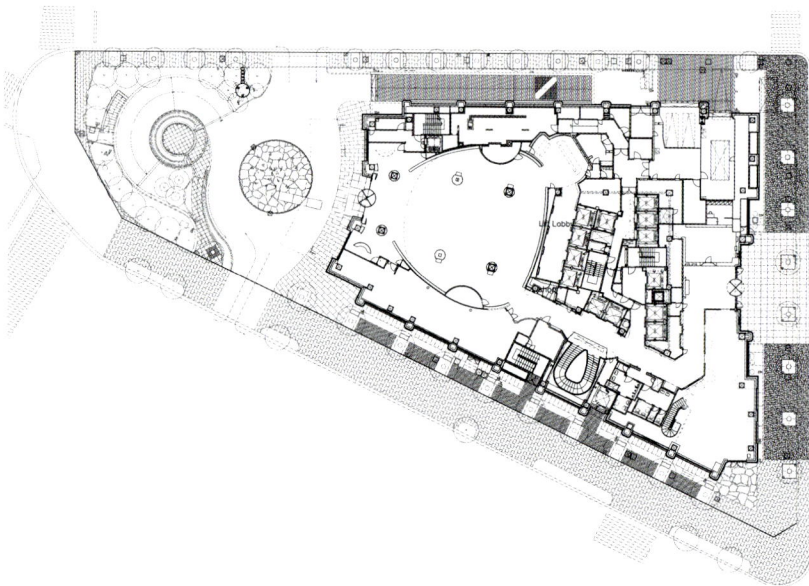

Local architecture firm Denton Corker Marshall chose a minimal design aesthetic that offers guests a sanctuary from the outside world. The care and attention lavished on its clientele are obvious in Jakarta's unconventional policies, such as the exclusive reservation for female guests or ample meeting space for savvy business travellers.

An Asian flair is evident in the 215 rooms, decorated in a lush yet soothing palette of dark red tones. Offering a pure and abstract design experience, Alila Jakarta is an exclusive and surprisingly isolated place to replenish your energy reserves. Each of Alila Jakarta's rooms, suites and apartments features parquet floors, internet access and breathtaking views over Jakarta. All rooms are designed with modern architecture, stylish simplicity and convenience.

这家雅加达当地的设计公司选择了简约的装饰风格来彰显典雅，使艾里拉酒店成为这一喧嚣区域中的"避风港"。酒店因其非同寻常的做法而显得别具一格，比如说只接待女性客人、为见多识广的商业旅行者提供足够的会议空间等。

215间标准客房全部以深红色调为主，凸显亚洲风格。客房、套房以及公寓内全部采用镶花地板装饰，提供网络服务，融现代时尚和简约便捷于一身。

1. The minimalism style feels soothing
2. Guestroom in daytime
3. The warm and intimate lighting
4. Guestroom at night
5. The suite
6. The bedroom enjoys an aerial view of the city

1.简约风格给人安宁享受
2.日光下的客房
3.温馨的灯光设计
4.夜色中的客房
5.套房
6.窗外景色如画

Al Ronchetto

阿尔·罗恩查托酒店

Treviso, Italy
意大利 特雷维索

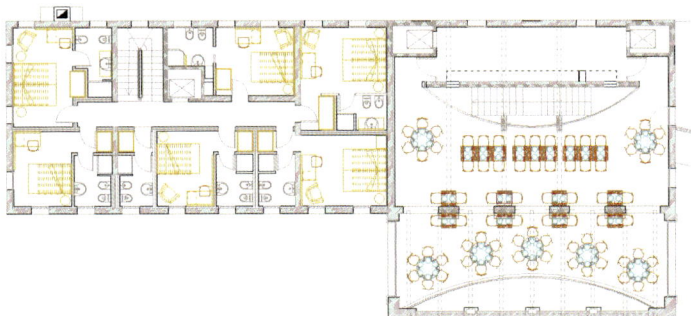

The Hotel Ristorante Al Ronchetto is located in Salgareda in the province of Treviso, quite close to the river Piave.

The hotel has 21 bedrooms created with special attention to detail and equipped with all of the services typical of contemporary hospitality. All of the rooms, characterised by ceilings with exposed beams, are furnished with wooden furniture. The bedrooms of the old house were planned in a more traditional style, whereas in the bedrooms in the expansion the design is more contemporary. The interior decoration plan offers a reinterpretation of the themes tied to the natural and archaic world of the Veneto countryside through a refined choice of finishing materials and decorative elements.

意大利餐厅酒店阿尔·罗恩查托位于特雷维索省，离皮亚韦河很近。

酒店里有21间客房是特别打造的，细部装饰很考究，配备了全套的服务设施，是典型的现代酒店风格。所有的客房都有一个特点，就是天花板上的横梁都是裸露在外的，而且都配备了实木家具。原来农舍里包含的那些客房采用了比较传统的风格，而扩建部分的客房则更具现代感。室内装饰是对威尼托区乡间的自然、古色古香的风格的重新诠释，设计师精心选择了考究的装饰材料和装饰元素。

1. The guestroom is natural, like the air in the countryside
2. The bathroom
3. The guestroom created with natural wood
4. The bathroom also has a countryside natural feeling

1.田园般恬静自然
2.浴室
3.全实木装饰的客房
4.浴室同样是农舍一样的纯朴自然

BALANCE holiday HOTEL

巴兰斯假日酒店

Zell am See, Austria

奥地利 滨湖采尔

Its inviting understated modern design not only generates an atmosphere of relaxing elegance, but is sensitively implemented to promote inner balance. The holistic concept behind the BALANCE holiday HOTEL aims to create a sense of "flowing privacy", a homogenous space that always places the individual at the centre.

The 47 unusually large guestrooms are in soothing natural wood and stone. The colour of each room is different. Every time coming into this hotel, you will have a fresh feeling. The atmosphere here is warm and happy. You can relax yourself from the busy daily life. If guests are feeling particularly invigorated and energised, the surrounding mountains, reaching an altitude of 2,743 metres, eagerly await them.

巴兰斯假日酒店用低调而现代的设计打造了高雅宜人的轻松氛围，同时巧妙地突出了室内的"平衡"。设计师秉承"流畅私密性"的理念，打造了统一的空间结构，使顾客成为中心。

47间超大号客房全部采用天然石材打造，温暖而亲切。每个房间都采用不同的颜色装饰，给客人新鲜感。暖意融融而又欢快愉悦的氛围让他们从喧嚣繁忙的城市生活中彻底"解脱"出来。如果客人在舒适的环境中备感经历充沛，那么就可以向四周海拔2743米高的山峰"进军"了。

1. The guestroom feels warm and homey
2. Panoramic view of the guestroom
3. Washroom in the guestroom
4. Washroom in the guestroom
5. Fireplace in the guestroom

1.温暖而亲切的客房
2.客房全景
3.客房卫生间
4.客房卫生间
5.客房中的壁炉

Becker's Hotel
贝克酒店

Trier, Germany
德国 特里尔

Becker's Hotel is located amidst the vineyards and rolling hills of Trier, one of the oldest cities in Europe, dating back to Roman times. Through the use of natural materials such as stone and wood, the property appears as if it almost nestles within its surroundings.

In the guestrooms, the stones give way to finely patterned dark wood floors combined with the glow of low spot lighting. A simple pane of glass separates the dark-tiled shower from the sleeping area, adding a hint of airiness to the otherwise earthy elements embodied in the hotel's design. At Becker's, German wine country's hospitality is writ large, but is certainly footnoted with a sophisticated, modern touch.

特里尔是欧洲最古老的城市之一，可以追溯到罗马时代。贝克酒店就是坐落在这座城市的葡萄园和绵延的山脉之间。通过对材料的运用（如石材和木材），这家酒店仿佛自然地"生长"在周围环境之中。

酒店客房里没有采用石材，而是用精心打磨的深色木质地板取而代之，结合低矮的聚光灯照明。卧室区与铺了深色瓷砖的淋浴区用一面玻璃屏风分隔开来，这样一来，原本有些粗俗的设计就增添了一抹空灵。德国的传统酒文化在贝克酒店里充分体现出来，不过当然是带有一种精雕细琢的现代韵味。

1. The dazzling black-and-white pattern
2. The suite
3. Black-and-white colour palette is a prevailing theme
4. The bathroom
5. A glimpse of the bathroom

1.墙面黑白拼接的图案
2.套房
3.黑白色的设计主题
4.浴室
5.浴室一角

4

5

Central Palace Una Hotel
尤娜中央宫殿酒店

Catania, Italy
意大利 卡塔尼亚

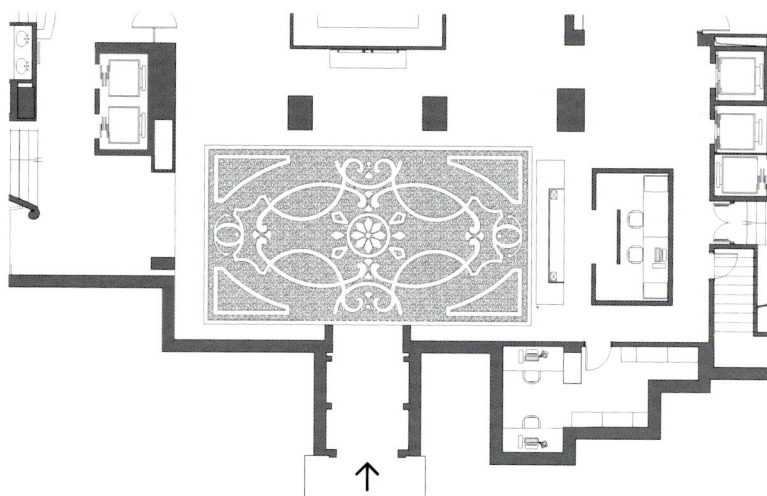

A prestigious building dating back to the early 1900s, in the artistic and commercial heart of Catania, was already a historical luxury hotel. The restoration project has been carried out in the view of maintaining the Sicilian style and traditional cultural heritage.

Rooms that give up onto "via Etnea" recover the topic of refined Mediterranean simplicity. The light comes to be the prevailing and essential element: white bed on an important wooden headboard embellished by a central majolica rosette. The interior design is based on ironic reinterpretations of classical furniture. Most indoor rooms allow the sight on a central patio that is treated as a Sicilian place.

这是一座古老的知名建筑，可以追溯到20世纪90年代初期，位于卡塔尼亚的艺术、商业中心地区，原本就是一家奢华酒店。此次翻修要保留原来的西西里风格和传统文化遗产。

朝向艾特街的客房呈现出考究的地中海式简洁风格。光成为这里的关键要素：白色的床采用木质床头板，床头板正中饰有跟楼梯风格一致的卡尔塔吉罗花饰陶器，是一朵玫瑰花的图案。室内设计是对古典主义家具有些嘲讽的全新诠释。大多数房间都能看到中央平台，平台也是按照西西里风格来处理的。

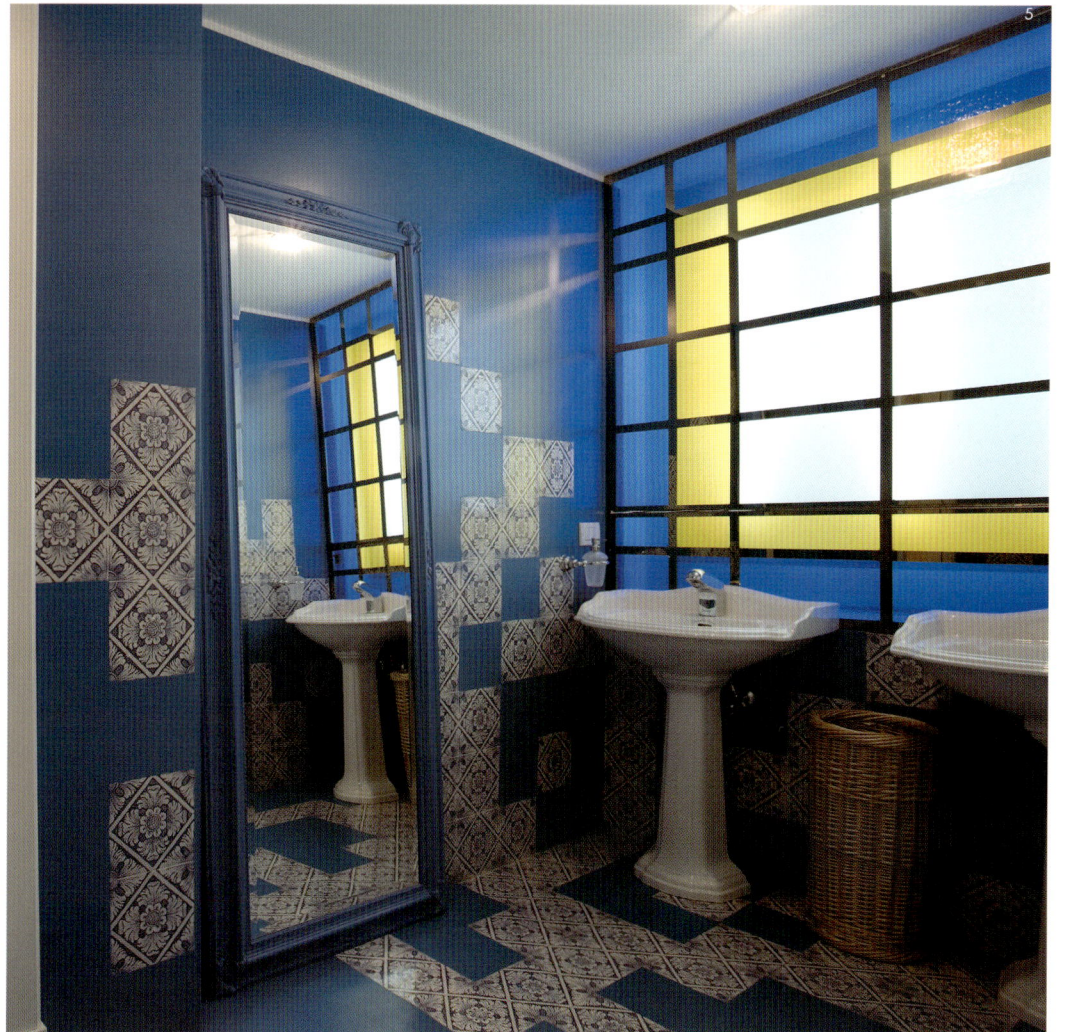

1. The lower level of the duplex guestroom
2. The duplex guestroom
3. The top level of the duplex guestroom
4. The guestroom in daytime
5. The washroom design style is typical of a holiday resort

1.跃层客房下层
2.跃层客房
3.跃层客房上层
4.阳光下的客房
5.假日风情的卫生间设计

Design Boutique Hotel Sun House
太阳屋精品设计酒店

Banovci, Slovenia
斯洛文尼亚 巴诺西

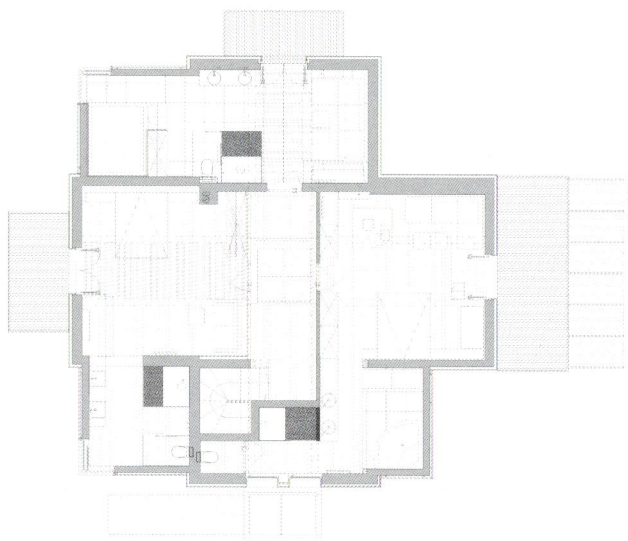

The thematic guest suites are situated on both the ground and the first floor. They are accessible from the common spaces. Each and every suite has its own story, providing this small boutique hotel with some extra charm and personal note.

The materials are quite agreeable and a lot of natural wood was used too. The colours used on the walls are mostly in warm brown tones. In opposition to the usual practice in the hotel rooms, this time the bathrooms and toilets are designed as a part of the entire space and marked off only with translucent or semi-translucent glass walls. An essential part of the suites' integrated image is graphic design, presented either as graphic pieces on walls or as printed fabrics of pillows and beddings.

主题套房位于一楼和二楼，从公共空间就能到达。每间套房都有自己的特色，这使得这家规模不大的精品酒店别具风韵，富于个性。

材料的选择非常亲切宜人，并且用了大量的天然木材。墙面上采用的色彩主要是温暖的棕色色调。卫浴空间的设计没有遵循客房设计的一贯做法，设计师将这一空间设计为整个客房空间的一部分，仅用透明或者半透明的玻璃墙隔开。套房的一大亮点是其平面图案设计，或者体现在墙上，或者印在枕头等床上用品上。

1. The green painting on the wall is a key element in the room, bringing vigour and liveliness to the space
2. The staircase links the living room on the ground floor and the bedroom on the first floor
3. View of the living room from the bathroom with transparent glass walls
4. The simple and clear guestroom
5. The spacious room with an illusionary decoration on the bedside
6. Wood continues from the floor to the wall
7. The glass-wall bathroom with glowing mosaic tiles

1.白色中点缀的绿色让房间生机盎然，床头的手绘墙画是亮点
2.楼梯连接一层起居室和二层卧室
3.玻璃浴室和起居室
4.简约大方的客房设计
5.室内空间开阔，床头装饰带有梦幻色彩
6.实木从地板延伸至墙面
7.玻璃浴室，马赛克墙面流光溢彩

1

Distrito Capital
首都地区商务酒店

Mexico City, Mexico
墨西哥 墨西哥城

Distrito Capital is located in the Santa Fe neighbourhood in Mexico City, one hour from the airport. Surprising interiors, dazzling panoramic views and double-height ceilings are a few of the eye-catching highlights of Distrito Capital. Located in the highest area of Mexico City – the skyscraper district of Santa Fe – this hotel is a testament to how cool Mexico's capital has become in recent years.

Designed around the idea of creative minimalism, the 30 well-appointed guestrooms and suites look more like chic art spaces than hotel rooms. Any visitor will be simultaneously awed by impeccable design touches and excited by personal service flourishes. Fashionable without being zeitgeisty, the inviting décor allows visitors to truly kick back and relax. Guests will feel like they've stepped into their dream apartment.

首都地区商务酒店位于墨西哥城Santa Fe区，距机场仅有一小时车程。内部令人赞叹的室内装潢，双层高的天花板和外部令人目眩的景色是整个酒店的亮点。这家酒店是墨西哥城近些年来渐趋时尚的一个证明。

酒店的设计围绕着具有创造性的简约风格构思展开，30间设施齐备的客房和套房看起来更像是别致的艺术空间。任何客人都会被这里无懈可击的设计和令人兴奋的个性化服务所打动。客人们来到这里会感觉来到了自己的梦中公寓。

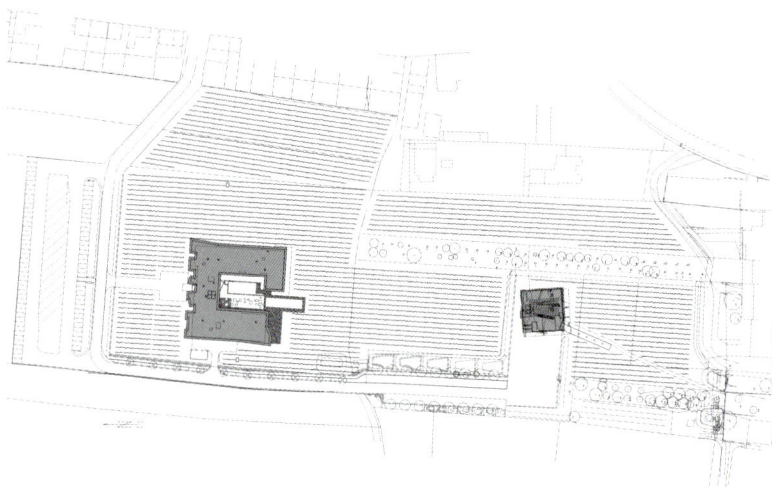

1. The spacious guestroom
2. Furnishings in the guestroom
3. Spectacular view
4. Innovative guestroom design
5. Office space in the guestroom

1.客房空间开阔
2.客房家具陈设
3.视野开阔
4.创造性的客房设计
5.客房中办公空间

3
4

Dream Hotel

梦境酒店

Florence, Italy
意大利 佛罗伦萨

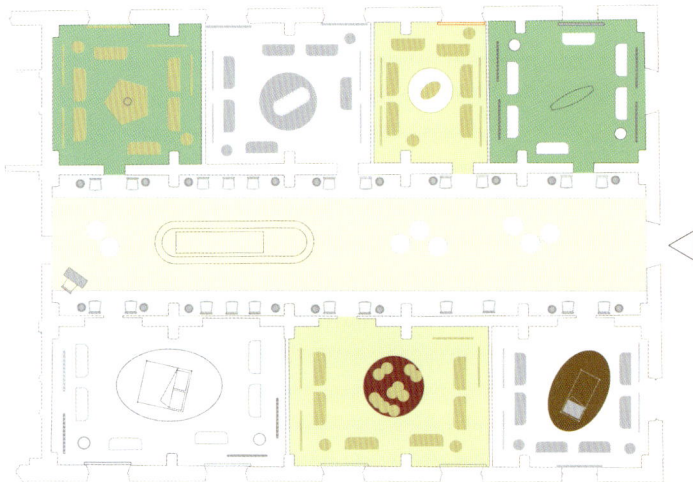

The hotel is located in the well-known capital of art – Florence. Correspondingly, the design of the hotel is full of an artistic air. The designers aim to create a hotel that is extremely artistic and dreamy, to continue visitors' experiences in the city.

The areas of every single exhibitor are transformed into delicate bedrooms characterised by different interpretations of dreamy spaces: a hammock, a romantic tester-bed, a golden cradle, a "thousand and a night" room furnished with lots of pillows, a poetic tree-house. Bed heads, chairs and small tables in wrought iron, covered with charming sheets of tulle, complete the etheric exhibition and become innovative displays to present the latest creations of the New Beat(s) area of Pitti Uomo fair.

梦境酒店坐落在著名的艺术之都佛罗伦萨，酒店的设计也契合这座城市的艺术气息。设计师试图通过设计手段，呈现出一个艺术化的、梦幻的酒店，延续这座城市带给游客的艺术感受。

每一个展区都变成了精致的卧室，设计中运用各种元素诠释梦幻空间：一张吊床、一张散发浪漫气息的试验床、金色的摇篮、诗意十足的木屋。床头、椅子以及小桌子全部采用熟铁打造而成，上面铺设着柔软的薄纱，从而打造了完美的展示空间。

1. A dreamy space with soft sheets of tulle
2. Softness spreads to every corner
3. The exquisite guestroom looks like an exhibition hall
4. The sheets of tulle contribute to the dreamy effect

1.轻纱环绕的梦境
2.每个角落都是轻柔的梦幻
3.别致的客房如同展示空间
4.幔纱营造梦幻氛围

Epic Hotel & Residences
史诗酒店

Miami, USA
美国 迈阿密

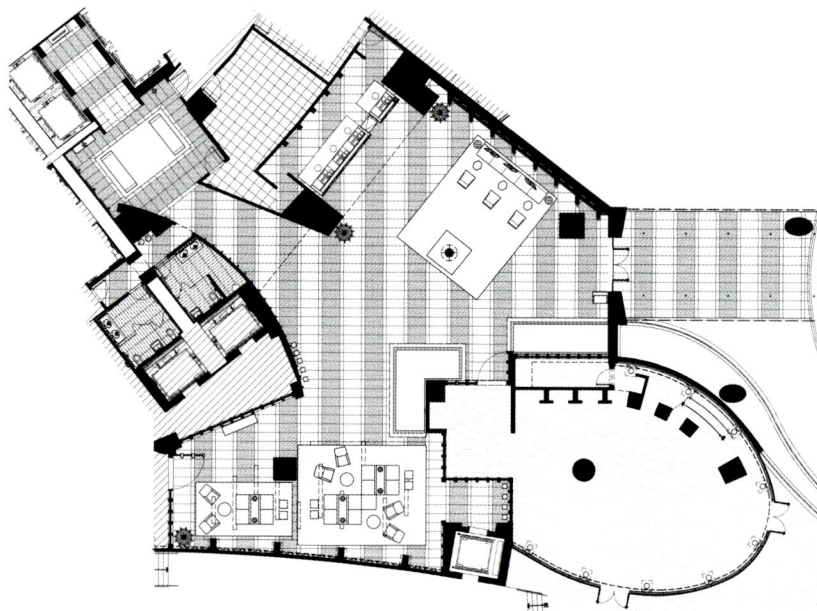

Epic Hotel contradicts the common depiction of the bright white and pastel-coloured Miami and presents a sophisticated, cultivated version of the city for the international traveller. Warmth, graceful simplicity and understated elegance are brought forth through the adept use of texture and light throughout the space. A neutral palette, natural textures and clean lines create a relaxed urban resort for an international, well-travelled crowd.

The Presidential Suite Parlour contains contemporary furniture in warm woods and cream lacquer. Lighting elements include a slivered coconut shell dining pendant and a sculptural pleated silk console lamp.

史诗酒店跟迈阿密常见的明亮的清淡柔和色彩不同，它展现给国际旅客的是一个精致、文雅的形象。温暖、优雅、简洁、低调的雅致，这样的空间风格通过巧妙地利用材料的质地与照明打造出来。素雅的色彩、自然的材质、简洁的线条，这些元素为有丰富旅游经验的国际旅客打造出一个轻松的城市休闲度假酒店。

总统套房大厅里有现代的家具，温暖的木质材料，涂着奶油色的漆。照明元素包括用餐区的一个椰子壳吊灯和一个雕塑般的有褶的丝绸悬臂吊灯。

1. The comfortable guestroom
2. The surrounding one hundred and eighty degree ocean views
3. Detail of the headboard

1.舒适的客房
2.180° 的海洋景观视野
3.床头细部

Fortress Hotel

堡垒酒店

Sri Lanka
斯里兰卡

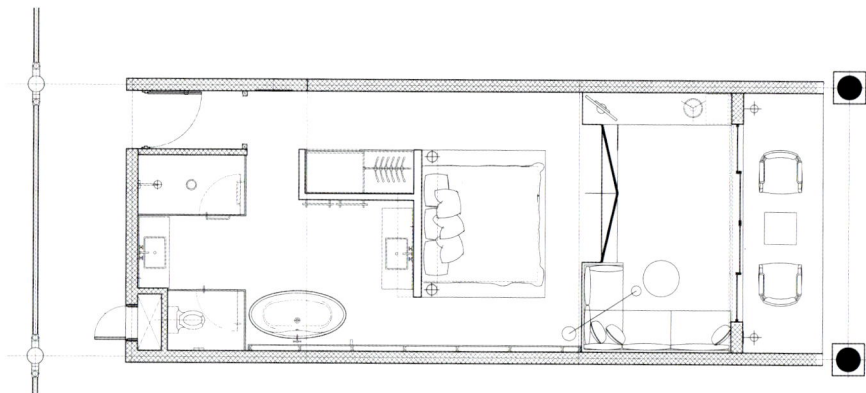

Fortress was a challenge to C&C Studio because it was originally designed architecturally as a three-star budget hotel. They were required to turn this typical modular room into something never foreseen and to a market of users who expected something extraordinary. This situation really made them focus much more on the details than over-all layouts, as the room size, envelope and structure were fixed.

The guestrooms are redesigned to be novel and modern, with simple and brightly-coloured furniture. The different areas, for example, the bedroom area and living room area, are defined by height difference. On the circular carpet, the curved sofa and the armchairs are set around the coffee table, creating a warm and intimate ambience.

堡垒酒店对C&C工作室来说是个挑战，因为这家酒店从建筑设计的角度来说本是一家3星级酒店。客户要求C&C工作室将这些标准间打造成以前从未见过的客房，目标群体是那些期待非同凡响的酒店体验的客人。这使得他们更关注细部设计，而不是整体布局，因为房间的大小和结构都已经固定了。

客房布局打破常规，新颖而富有现代感，家具造型简约，色彩鲜明。客房内的地平有高差，地台上或设计成卧室，或布置起居室，圆形地毯上，弧形的沙发和扶手椅围着圆形茶几，有家人聚餐的温馨感。

2

3

1. The living room area placed on a higher stage
2. The simple yet unique double room
3. The bathroom
4. The lighting for the bed creates a warm atmosphere
5. The bathroom
6. The living room is suitable for family reunion

1.布置在地台上的起居室
2.简约而独特的双人间
3.浴室
4.床头灯光温暖
5.浴室
6.适于家人围坐的起居室

Gabriel Hotel

加布里埃尔酒店

Paris, France

法国 巴黎

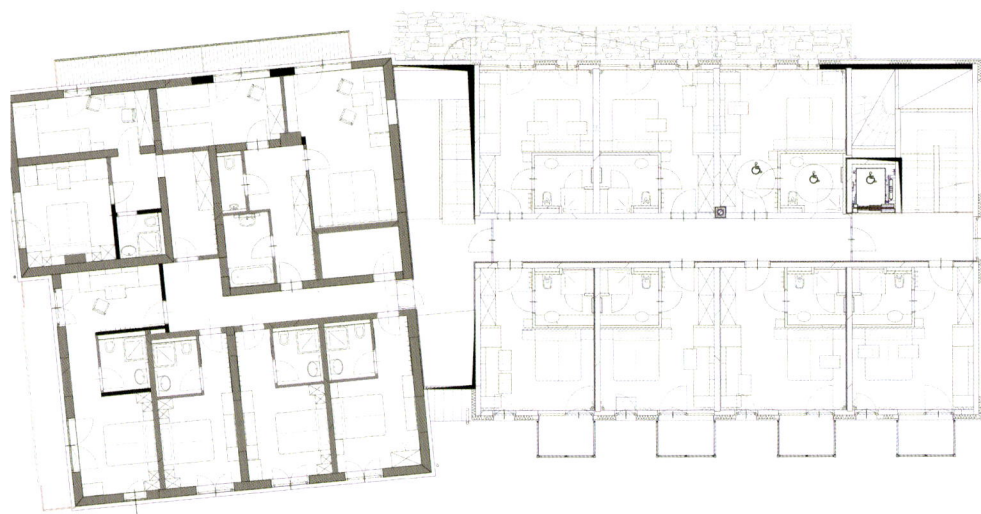

The hotel's Glowing Rooms and Suites provide their inhabitants with a variety of light and sound programmes that correspond to the natural stages of the human sleep cycle, giving users a peaceful and invigorating night. Gently curving walls and LED back-lit sliding doors assure mellow light and soft transitions. Guestroom interiors celebrate Zen minimalist charm in white and pastel. Linear furniture with cruise-liner portholes and curves creates a harmonious balance best expressed in the guestroom centrepiece, a vertical "cabinet" comprised of a mirror-hidden television, a folding desk and a mini bar. Signature ballerinas throughout the property add a whimsical touch to the dreamy 1930s feel.

加布里埃尔酒店的"光辉"客房和套房通过变幻的光与声来模仿人类睡眠的各个阶段，让客人享受宁静、酣畅的夜晚。微呈曲线的墙壁和采用LED背光照明的滑动门保证了室内温和的光线和柔和的过渡。客房的室内设计呈现出禅意十足的极简主义风格，色彩上清淡柔和。几何线条的家具上面带有游轮舷窗和曲线图案，创造出一种和谐的美感，而这种美感又在客房的焦点——一个竖向"陈列室"——上体现出来，里面有隐藏在镜子中的电视、一张折叠桌和一个迷你酒吧。芭蕾舞女演员的形象遍布酒店各处，带来一种梦幻般的20世纪30年代的感觉。

1. The colourful shutter is adopted in the white guestroom
2. Natural light is combined with artificial lighting
3. The partition with golden patterns
4. Signature ballerinas
5. The washroom
6. The private SPA
7. A TV set and a folding desk are smartly integrated in the "cabinet"

1.白色的客房装饰彩色百叶窗
2.自然光与人工照明相交织
3.金色点缀的隔断
4.柜子上的图案充满趣味
5.卫生间
6.私人SPA
7.电视墙与置物台相结合，巧妙利用空间

Gastwerk Hotel

房客之家酒店

Hamburg, Germany
德国 汉堡

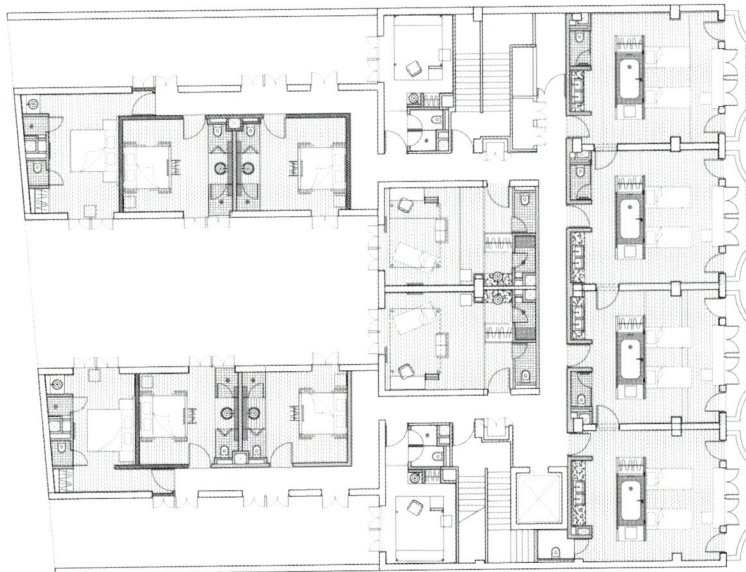

Seasoned hotelier Kai Hollmann turned Hamburg's 19th-century municipal gasworks ("Gaswerk") into Europe's first loft-style hotel as easily as dropping a "t" into its name, instantly transforming the complex into a "guest works".

The "old" rooms feature original brick walls and arched multipane windows that can be darkened with sliding panels of felt. Finest walnut wood trimmings, tall casement windows and unmasked concrete walls garnish "new" conference rooms and penthouse suites. Rotating propeller-headed ceiling fans pay homage to the many mechanical devices that once whirred away in the power station.

酒店老板卡伊·霍夫曼将汉堡这家19世纪的市政所属的煤气厂（德语为"Gaswerk"）变为欧洲第一家阁楼式酒店。转变很简单，就是在其德语原名中加入一个字母"t"，就把它变为了"房客之家"（"Gastwerk"在德语中有"房客"之意）。

"老"房间的特点是保留原来的砖墙和小块玻璃拼接的拱形窗户，窗户配有滑动毛毡板条来遮光。最好的胡桃木装饰、高高的窗扉以及不去隐藏的水泥墙打造出"新"会议室和楼顶套房。天花板上的螺旋桨式风扇充满了对曾经在这座发电站中运转的多种机械装置的敬意。

1. The peculiar beauty of the brick wall
2. Living room in the attic suite
3. The attic suite
4. Sitting area in the corridor
5. The working area

1.砖墙创造特殊美感
2.顶楼套房的起居空间
3.顶楼套房
4.走廊里的休息处
5.办公空间

Golden Sands Resort by Shangri-La, Penang

槟榔屿香格里拉金沙大酒店

Penang, Malaysia
马来西亚 槟榔屿

Golden Sands Resort by Shangri-La, located on Penang's popular Batu Feringgi beach, is a veritable tropical paradise for vacationers and families.

All the guestrooms are semi-open, each enjoying at least one open-air terrace. Guests could enjoy comfortable breezes in the guestroom. The white-toned guestrooms are decorated with bright-colour patterns, fully demonstrating the tropical warmth and romance.

香格里拉金沙大酒店位于槟榔屿有名的都丁宜海滩，对于度假或者全家出游来说是一个名符其实的热带天堂。

客房全部是半开敞设计，每个房间至少有一个可以享受户外新鲜空气的露台，海风吹入客房中，给客人美妙的享受。客房设计以白色为基调，带有色彩艳丽的图案作装饰，尽显热带的热情和浪漫。

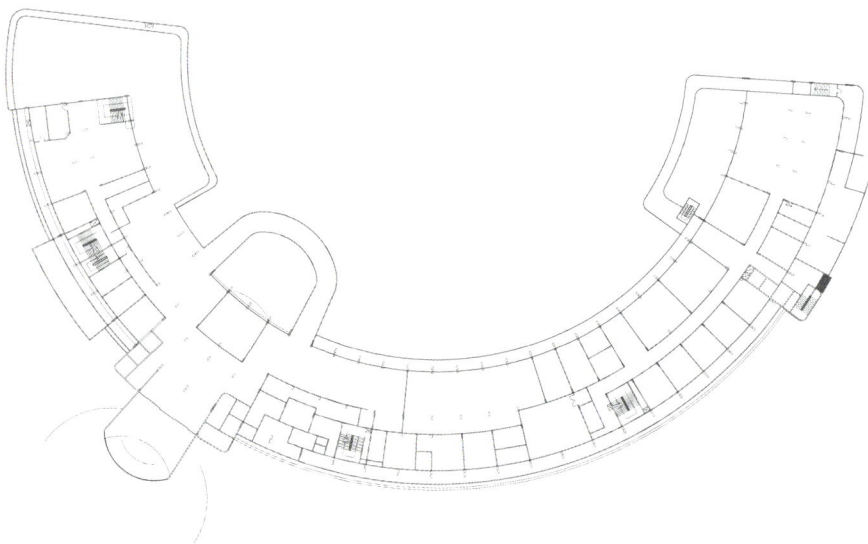

1. Family suite
2. You could enjoy the cool breezes when in bed
3. The scenery of the sea outside is part of the interior decoration

1.家庭间
2.在床上就能感受海风的吹拂
3.海景融入客房装饰之中

Hard Rock Hotel
硬岩酒店

Las Vegas, USA
美国 拉斯维加斯

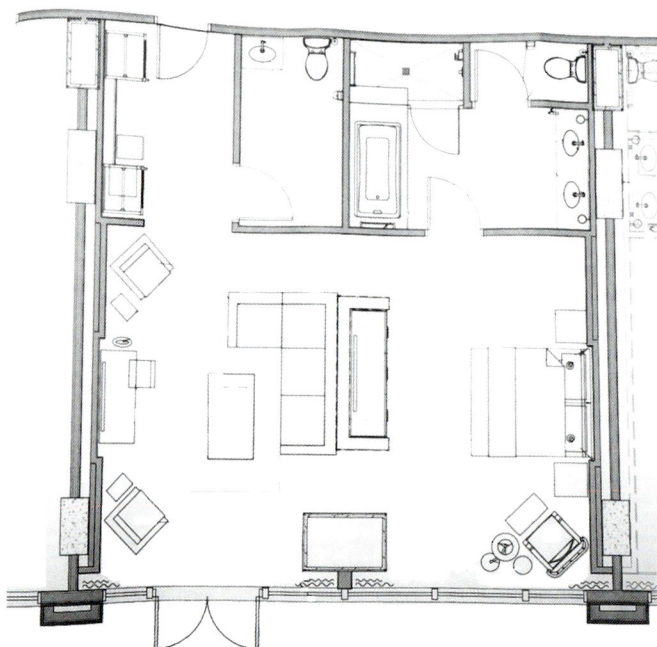

From empty concrete shell, the complete budget of design, construction and decorating was $250,000. Construction took place from February 2009 to June 2009.

The design concept of this suite was to create a dark twisted take on a library/cigar bar. The design features Wenge wall panelling cut into custom shapes and divided with silver, metal striping. The metallic-silver porcelain wall tile contrasts with the rich hardwood floors and gives the suite a modern feel – but the bronze-mirrored ceiling, croc-skin custom furniture, and giant hand-painted wall murals of girls and gorilla make this a distinctly dark, savage, sexy urban jungle.

建筑起初只有混凝土构架，从设计、施工到装修，总预算是25万美元。施工从2009年2月开始，于6月结束。

大猩猩套房的设计理念是打造阅读、吸烟室一般的奇异的暗色空间。设计以阁壁为特色，墙面上的镶板是特别设计的造型，有银色的金属条纹。这种银色金属质感的瓷砖墙面跟温暖的硬木地板形成对比，给套房内带来一丝现代气息，但是青铜色的反光天花板、鳄鱼皮表面的定制家具、巨大的手绘壁画（画上是女孩与大猩猩），这些元素又将套房打造成阴暗、蒙昧、感性的城市丛林。

The intent of this design was to create a party suite with a modern, airy, Miami feel. The primary feature is a giant C-shaped bed lined with blue, imported Italian tile. LED rope lighting and mirrored toe-kicks beneath the bed create the illusion that the bed floats above the floor. Laid out like a party lounge, a solid-surface DJ table in "Milk Glass" foregrounds the second most noticeable feature: a wall of turntables and jam boxes framed in LED-lit alcoves. Additional features include custom-designed minimalist furniture, a custom, colour-matched rug produced in Sweden, floor-to-ceiling sheers, sconce lighting and mirrored ceiling niches. The primary challenge was a demanding timeline, as they had been hired late in the construction process.

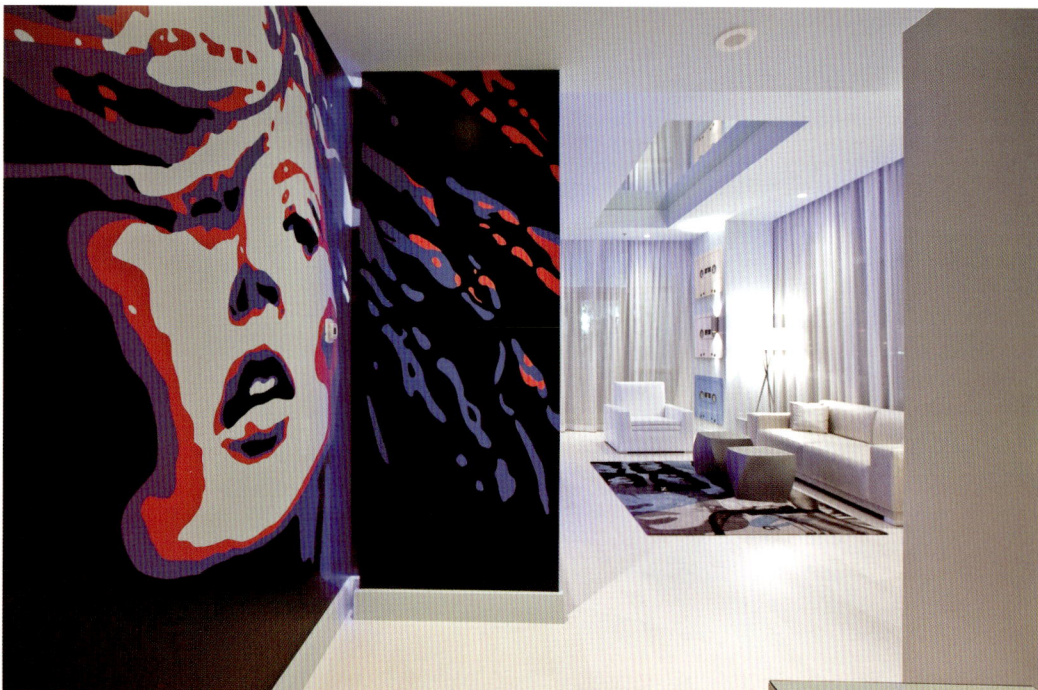

迈阿密蓝色套房的设计理念是打造现代、通透的迈阿密风情套房。套房最突出的特点就是"C"形的大床，装饰着意大利进口的蓝色瓷砖。LED彩虹灯和床下的反光材料让人产生一种幻觉，觉得床仿佛漂浮在空中。套房的布局就像休息室，一张表面坚实的乳白色玻璃的DJ专用桌成为了第二个突出特点的前景；第二个特点就是：一面由唱机的转盘和果酱盒子组成的墙，墙的周围有LED照明的壁龛。其他特点还包括：定制的极简抽象家具、从瑞典定制的颜色巧妙搭配的地毯、落地窗、壁突式烛台照明、反光天花板。由于直到施工接近尾声时才接到设计任务，设计师面临的最大挑战就是工期太短。

The intent of the graffiti room was to create a London-style, punk rock party suite. Bands of recessed mirrors and hand-painted lines in every direction, hand-painted wall murals, and spot lighting on the walls and ceiling pack the room with movement and action.

Stone-like grey wall tiles, black hardwood floors, custom modern furniture and a custom, handmade Union Jack rug complete the London underground feel.

这家酒店里的涂鸦艺术空间的设计理念是打造伦敦风情的朋克摇滚套房。凹陷的镜子、手绘的涂鸦线条、手绘壁画以及墙上和天花板上的投光灯，都让房间内充满了动感。

灰色墙砖、黑色硬木地板、定制的现代家具以及定制的手工缝制的英国国旗图案的地毯，共同打造出伦敦地铁的感觉。

The concept of this design began with the concept of creating a hotel room with the illusion of a two-storey tree line in the centre. Starting from this feature, the hand-picked colours of tree branches were the basis for the colour scheme for the rest of the room. Hardwood floors complement the nature motif. Cream walls of textured tiles anchor each side of the room. To create contrast and clarify that this space is still a party suite, a rich blue band spanning from wall, to ceiling, to wall, divides the room down the centre. The blue mural features hand-painted smoke in dark blue, leaves and skulls in metallic gold, and LED rope lighting to illuminate the art.

树线套房的设计理念是打造这样一种酒店客房：使人在里面会产生一种幻觉，觉得房间中间好像有两层高的树线。从这个特点出发，房间内其他地方的色彩基调就定下来了：精选的树枝颜色。硬木地板强化了自然的主题。奶油色的墙壁上铺着质地优良的瓷砖。为了对比、强调出这个空间是套房，特别设计了一条蓝色的色带，从墙到天花板再到墙，从中间把这个房间一分为二。蓝色的壁画上是手绘的深蓝色的烟和金色的树叶和骷髅，LED彩虹灯照明进一步渲染了其艺术性。

Haymarket Hotel

海玛特酒店

London, UK
英国 伦敦

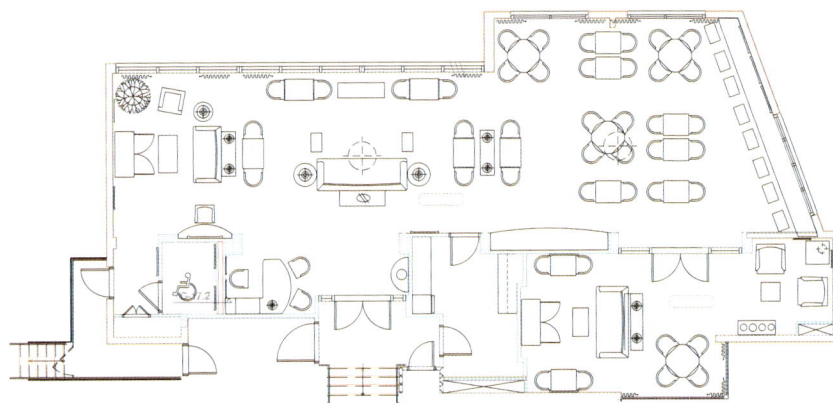

Haymarket Hotel is situated off London's Haymarket in the heart of the theatre district and right next to the Haymarket Theatre Royal and just steps from Trafalgar Square, the National Gallery and St. James's Park. A bold step away from cookie-cutter minimalism, Haymarket Hotel fuses contemporary and classical references in an ultra-central London location.

Guestrooms are individually furnished and feature custom-made pieces, avoiding the "designer formula" look and emphasising rich texture and colour. The colour palette and the choice of patterns are both aimed at the creation of enchanting amenities.

海玛特酒店位于伦敦海玛特剧院区的中心地带，紧临海玛特皇家剧院，与特拉法加广场、国家艺术馆和圣詹姆士公园仅几步之遥。海玛特酒店在设计上大胆走出最简单艺术派主义的固定模式，强调在伦敦超中心的位置上的现代和古典两种特色。

客房的布置按客户要求进行，带有个性化的特点，这样就能避免千篇一律的"设计师公式"产生的外观效果，同时强调丰富的质感和色彩。无论是颜色搭配还是图案的选择，设计师都意在创造一种自然舒适的环境。

1. The living room
2. The guestroom is full of natural air
3. The fabrics bring a countryside feeling
4. The reminiscent cabinet in the corner is a key element

1.起居室
2.充满自然气息
3.清新田园风
4.角落里怀旧的立柜平添趣味性

Hospes Madrid

马德里霍斯佩斯酒店

Madrid, Spain
西班牙 马德里

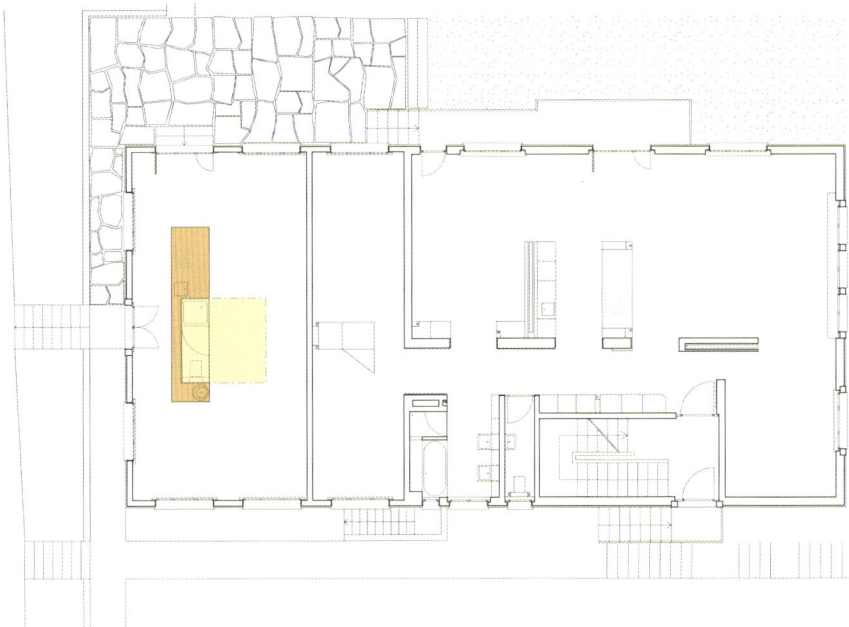

Originally an affluent apartment house with wrought-iron balconies designed in 1883 by architect José María de Aguilar, the handsome red-brick Hospes Madrid is an icon of Bourbon Restoration period architecture. So the Hospes Design Team needed to take special care that their modern additions were in harmony with the building's historic elements.

The narrow windows and the floor-to-ceiling curtains naturally bring a classical and noble air. In accordance, the designers adopted some classical style elements to create a space that is classical and modern at the same time. The facilities in the guestrooms make you feel as if you are in a palace, being pampered as a king or a queen.

这座建筑原本是一座公寓，由建筑师乔斯·玛利亚·阿圭勒于1883年设计，有着锻铁阳台和漂亮的红砖。这座"马德里霍斯佩斯"是波旁皇族复辟时期的一座代表性建筑。所以霍斯佩斯设计小组此次特别注重让现代元素跟这座建筑的历史元素能够彼此融合。

窄窗和落地窗帘给客房带来一丝古典而高贵的韵味，此外设计师还用到一些古典风格的设计元素，营造既高贵又现代的室内空间。客房内的设施也布置得如同宫殿一般，让入住的客人享受国王或王后一般的体验。

1. The different areas in the guestroom are compactly laid out
2. The sofas are particularly designed
3. Natural light is combined with artificial lighting
4. Every piece of the furnishings is carefully chosen
5. The bathroom
6. The living room

1.客房空间紧凑
2.沙发造型别致
3.自然光与人工照明相映成趣
4.每一件家居都是精心设计
5.浴室
6.起居室

5

6

Hotel Benkiraï

彭吉哈伊酒店

Saint-Tropez, France
法国 圣特罗佩

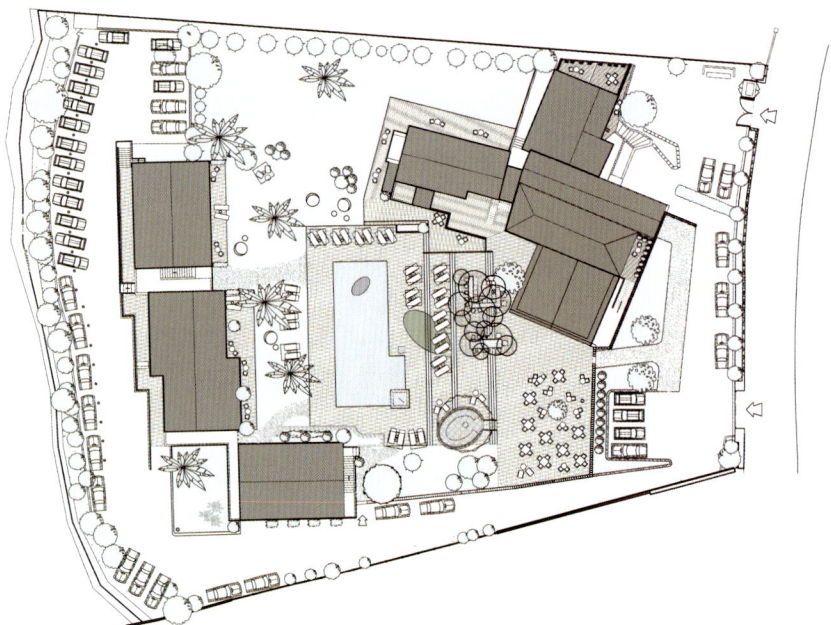

The selection of materials and the volumes produce a feeling of refined sensuality, harmoniously organised around the bed. The furniture is lacquered. Its pure lines are highlighted by the contrast with the concrete floor, natural grey in the bedrooms and blue in the bathrooms. There are no projecting handles; everything opens with a gentle touch of the finger. Each opening is located by a discrete stainless steel thumbprint. The bathroom opens onto the bedroom, separated by a satin curtain. The sink unit opens like a jewel box in front of a large suspended two-sided mirror, lit bathroom side, slightly smoked bedroom side.

材料的选择以及空间的大小使客房有一种考究的感官体验，围绕着床展开。家具都涂了漆。纯净的线条通过与水泥楼板、卧室中自然的灰色和浴室中的蓝色的对比显得更加突出。没有突出出来的把手，每样东西手指轻轻一碰都可以打开。每个开口处都设有各不相同的不锈钢个性特征。浴室朝向卧室开门，用一个缎面帷幕隔开。洗手盆就像珠宝箱一样，前面是一面巨大的悬垂双面镜子，在浴室一边是照亮的，而卧室一边则较暗，微微呈烟熏色。

1. The ceiling design is eye-catching
2. The open lavabo
3. The pretty terrace outside of the guestroom
4. The peculiar "bird-nest" lamp on the wall
5. The pretty guestroom terrace
6. The bathroom

1.天花板的设计别致有趣
2.开放的洗手台
3.客房附带小露台
4."鸟巢"壁灯别具一格
5.客房露台
6.浴室

5

6

7. The blue and white pattern creates a quiet
space, just like the house of a hermit

8. The pink-and-purple colour palette is as sweet
as a lovely girl

9. The retro furniture and furnishings

7.蓝白色花纹的图案如隐者居住般清幽

8.粉紫色装饰如少女般甜美

9.复古的家居和陈设

Hotel EOS
EOS酒店

Lecce, Italy
意大利 拉察

The interior design follows "art hotel" concept and different designers have developed hotel room theme involving local furniture companies. In the rooms, Salento and design merge in a free interpretation of style; the designers propose contained dimensions that rarely, and in any case only slightly, exceed the standard minimum sizes. Freed from useless encumbrances, the rooms were planned on the basis of the real requirements of the modern traveller favouring functionality and comfort, and they are distinguished by the diversification of their furnishings. The bathrooms are proportionately more spacious and in harmony with the décor; they have more facilities and comforts than the traditional three-star hotel, such as a roomy shower or a direct telephone line; they have regular contours, a dynamic style and a captivating design, just like the thirty rooms, which are all different from each other.

室内遵循"艺术酒店"的设计理念，客房由不同的设计师打造，各自具有独特的风格。房间设计在面积上仅比标准小型客房小一点，摒弃了过多的装饰，只要满足现代旅行者对功能及舒适度的要求即可。浴室空间相对开阔一些，装饰与整体格调一致。值得一提的是，这里配备的设施远比传统的三星级酒店多样化，包括淋浴、外线电话等。浴室外观整洁，风格充满活力，极具魅力的设计正如30间彼此各不相同的客房一样。

1. The bedside lighting and wall painting
2. The design of the room is simple yet artistic
3. The curve wall is quite playful
4. The decoration on the wall completes an artistic space
5. The curve wall is quite playful

1.床头的灯光和墙画设计
2.房间设计简约而不乏艺术感
3.弧形墙面设计充满趣味
4.墙壁装饰让房间充满诗意
5.弧形墙面设计充满趣味

1. The adornments on the headboard look like the badge of a ship
2. The Moroccan antiques complete the interior design
3. Dining room in the suite
4. The washroom

1.床头装饰如同海船上的徽章
2.摩洛哥特色的室内装饰
3.套房餐厅
4.卫生间

Hotel Palomar Arlington

阿灵顿帕洛玛酒店

Arlington, Virginia, USA

美国 弗吉尼亚州 阿灵顿

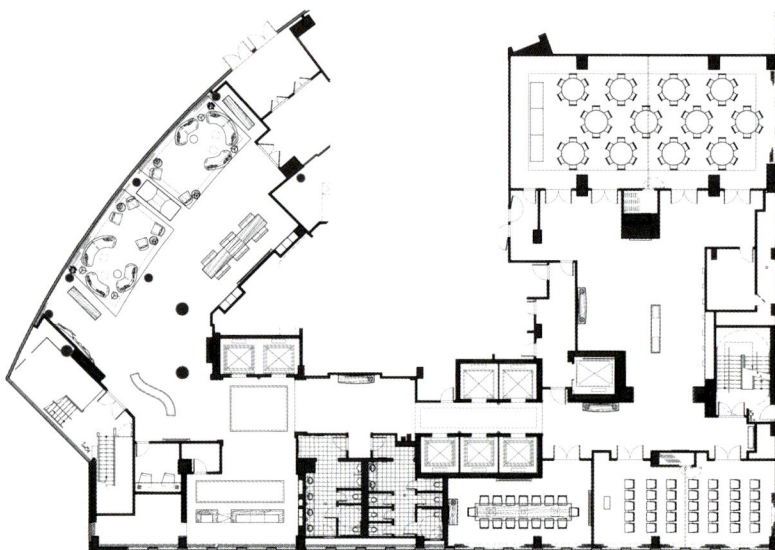

Taking inspiration from the hotel's very location – a striking glass-and-steel tower rising from the banks of a historic river, the design team created the sophisticated, inviting interiors of this four-star hotel based on contrast and balance – earth and water, modern and traditional, feminine and masculine.

The guestrooms are designed with a countryside style. The fabric sofas with delicate patterns are carefully selected, well corresponding with the patterns on the carpet. Picturesque views are available outside the window. The big red decorations in the guestrooms are quite eye-catching, bringing some vigour to the spaces. At night, the exquisite lamps would give out colourful light, creating an intimate atmosphere.

设计师从酒店的地理位置上获取灵感——这家酒店是坐落在一条历史悠久的河边的一座玻璃加钢铁的高层建筑，十分引人注目。于是设计师相应地为这家四星级酒店设计了丰富、宜人的室内空间，设计理念是土与水、现代与传统、阴柔与阳刚的对比与平衡。

设计师给客人带来一系列如同田园假日的客房设计。精心挑选的布艺沙发，上面有精致的花纹，与地毯的花纹相映成趣。窗外风景如画。高大的红色装饰为客房中增添了活跃的色彩。夜晚，精美的灯饰散发出各色的灯光照在房间里，十分温馨。

1. Picturesque views are available
2. The decoration of the guestroom evokes a romantic holiday air
3. Living room in the suite
4. Washroom in the guestroom
5. The suite

1.窗外风景如画
2.客房装饰富有假日浪漫色彩
3.套房起居室
4.客房卫生间
5.套房

Hotel Palomar Washington

华盛顿帕洛马酒店

Washington, USA
美国 华盛顿

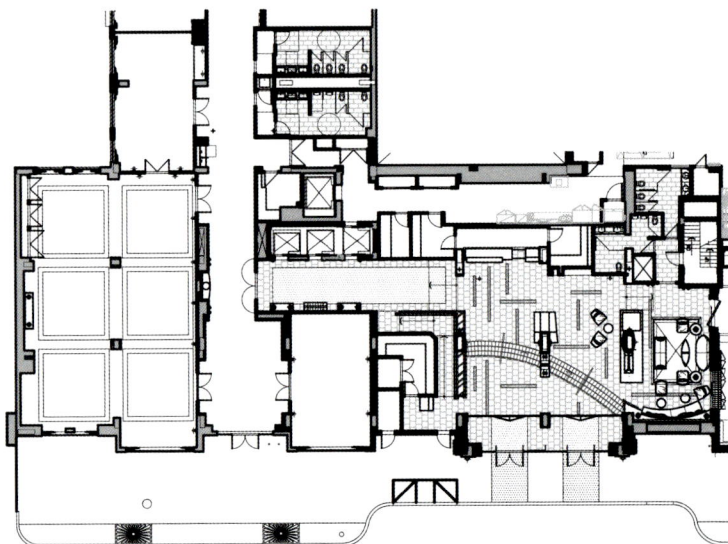

You can expect the unexpected at the Palomar Washington, where no detail is too minuscule. Striking lacquer finishes, exclusive materials and geometric forms are reminiscent throughout this property.

This soothing colour palette leads to the guestroom where a chic bronze-and-sable alligator pattern carpet is juxtaposed against a taupe wallcovering with threads of slightly shimmering gold horizontal lines. The superior "Kimpton" bed with imported crisp white bedding is anchored by a full-height upholstered headboard with enhanced tufted stitching. At the full-height window wall hangs a lustrous interlocking circle motif drapery. Featured bedside is the "piece de resistance": a lighted faux alabaster nightstand.

在华盛顿帕洛马酒店，你可以期待意料之外的惊喜。酒店中醒目的油漆处理、独特的材料以及几何形态比比皆是，有一种怀旧之感。

这样的色调令人备感抚慰，一直延伸到客房，这里，地上铺着漂亮的红褐色黑貂鳄鱼花纹的地毯，墙上，灰褐色的墙面涂料上面有微微发光的水平金线。超凡的金普顿大床上是整洁的进口床上用品，软垫床头板一直延伸到天花板，上面还有穗饰。窗帘从顶端垂下，上面有光泽的连锁圆圈图案。床边是个焦点——人造条纹大理石床头柜。

1. The nightstand is a key element
2. The comfortable guestroom would help get rid of your fatigue
3. Washroom in the guestroom

1.床头柜设计是亮点
2.舒适的客房让人忘记疲倦
3.客房卫生间

Hotel Principe di Savoia

普林西比·迪萨瓦酒店

Milan, Italy
意大利 米兰

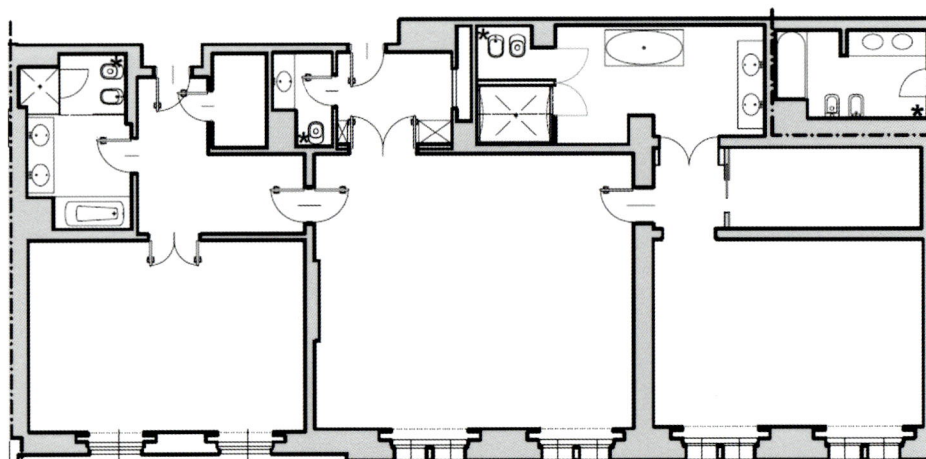

Hotel Principe di Savoia has undergone an exciting transformation. Nine new Principe Suites have been introduced and the Imperial Suite has undergone a complete transformation. The Principe Suites are appointed with hand-painted frescoes, traditional Italian furniture as well as deep purple armchairs and floor lamps creating a warm ambience. The large sitting rooms, part of each suite, are furnished with sumptuous sofas. The stunning bathrooms are spacious, light and equipped with a Lasa marble bath in the centre of the room as well as a shower adorned with striking handmade glass mosaics.

The redesign of the Imperial Suite combines a series of contemporary and classical elements. Striking paintings featuring interpretations of contemporary masterpieces have been specially created.

普林西比·迪萨瓦酒店进行了耳目一新的翻修。新增了9间普林西比套房，原来的皇家套房也经过了全面的翻新。普林西比套房内采用手绘壁画、传统的意大利家具、深紫色的扶手椅和地灯，营造了一种温暖舒适的氛围。宽敞的起居室是每间套房的一部分，配备了奢华的沙发。浴室宽敞、通透，中央是一个大理石浴盆，淋浴区采用手工制作的玻璃马赛克装饰。

皇家套房经过重新设计，结合了一系列现代与古典设计元素。设计师特别选取了几幅油画，对当代名作进行了诠释。

1. The luxurious Imperial Suite bedroom
2. The Imperial Suite living room, with the curtains hung from the ceiling, is just like an imperial palace
3. The Principe Suite bedroom looks like the bedroom of a princess
4. The Imperial Suite living room
5. The pattern design on the wall of the Imperial Suite bathroom is eye-catching
6. The Principe Suite bathroom

1.奢华的套房
2.从屋顶垂落的窗帘让套房如同宫殿
3.像小公主的卧室
4.套房起居室
5.浴室墙壁的花纹引人注目
6.浴室

Hotel Puerta America
美洲门户酒店

Madrid, Spain
西班牙 马德里

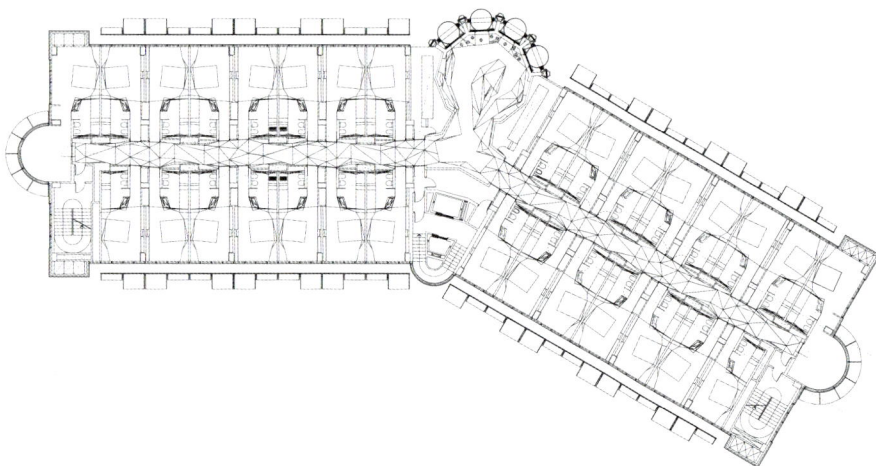

The room's entrance area is shaped as a funnel towards the view through the plate glass window. This includes a ramped floor of light grey rubber that finishes against the carpet in the same colour, introducing the user to the comfort of his bedroom. Like the edge of the visual cone from the entrance turning into a strip light and a fold in the ceiling, many more visual relationships reverberate through the geometry of the ceiling folds.

A long meandering sheet of stainless steel becomes the main organising element of the room: moving all along the back wall, it starts as desk, turns headrest and then seat. It "crashes" into the glass division wall and mutates into a bathtub and finally the shower.

客房的入口区设计成漏斗形，朝向玻璃窗。这个"漏斗"包括浅灰色的橡胶地面以及上面铺的同样颜色的地毯，将客人引向舒适的卧室。入口处的视锥的边缘逐渐转变成光线和天花板上的折线，这种视觉关联在折线形天花板的几何造型中无限反射。

整个房间里的一个贯穿始终的元素就是一个长长的蜿蜒不锈钢造型：沿着后墙，一开始是桌子，然后变成靠枕，然后是座椅，最后"撞进"玻璃分隔墙，变成浴缸，最后成了淋浴间。

2

3

1. The geometrical glass installation
2. Glass and mirror bring surprises to the space, representing break and recombination
3. Detail of the geometry
4. Simple lines and surfaces produce an unexpected effect
5. The irregular geometrical shapes seem casual, but actually carefully calculated

1.几何形状的玻璃装饰
2.玻璃和镜面，破裂和重组，给空间带来惊喜
3.几何形状细部
4.单纯的线与面的装饰，带来意想不到的效果
5.看似无规则的分割，其实都经过设计师的精心计算

Hotel Realm
王国酒店

Canberra, Australia
澳大利亚 堪培拉

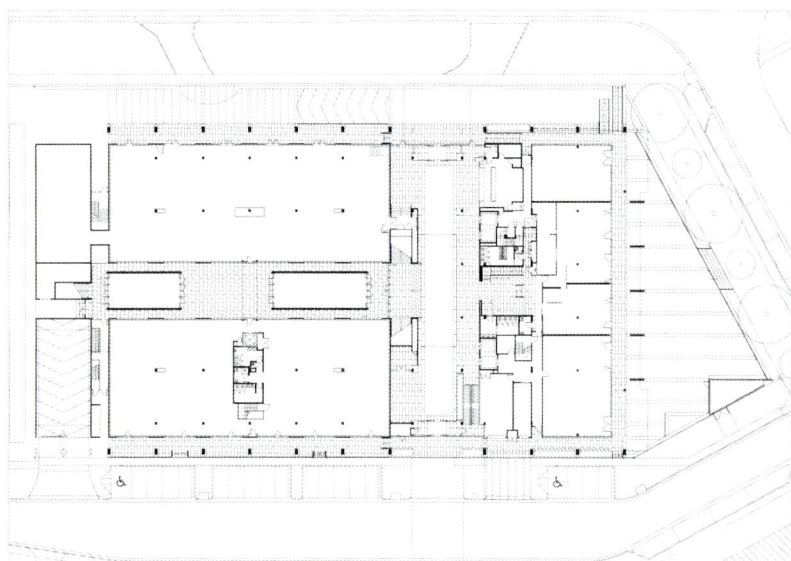

In a similar way the three building blocks which combine to make up the hotel building group also form the enclosure and edges to the major atrium at the heart of the hotel itself. In RAIA Jury's words, "Hotel Realm is commended for considerable control of material and proportion throughout the public and private accommodation spaces, resulting in consistently formal expression of wall planes and openings to the surrounding areas."

The guestrooms are carefully designed, from the spacious space, the spectacular view, to the comfortable furniture. They are aimed to bring the home-like feeling to guests. Neutral colours are adopted for the decoration and the furnishings in the guestrooms. The simple lines further enhanced the visual transparency of the space. Here guests are likely to relax both their bodies and their minds and have good rests.

酒店的三幢楼以相似的模式形成一个封闭空间，中心以一个中庭相连。用澳大利亚皇家建筑师学会评委的话说，"王国酒店在公共和私密空间的材料和比例都控制得很好，因此能在墙面和对周边区域的开口处有一致、正规的表现形式。"

设计师精心设计了客房，无论是开敞的空间、开阔的视野，还是舒适的家具，都试图给顾客带来家一般的享受。客房内的装饰和陈设都选择了中性色，简约的设计增加了视觉的通透感和开阔感，让顾客更容易放松身心，得到良好的休息。

1. The typical guestroom
2. The guestroom in sunset
3. The simple but comfortable guestroom

1.典型的客房
2.晚照下的客房
3.客房设计闲适简约

Hotel Rössli

罗思丽酒店

Bad Ragaz, Switzerland
瑞士 巴特拉加茨

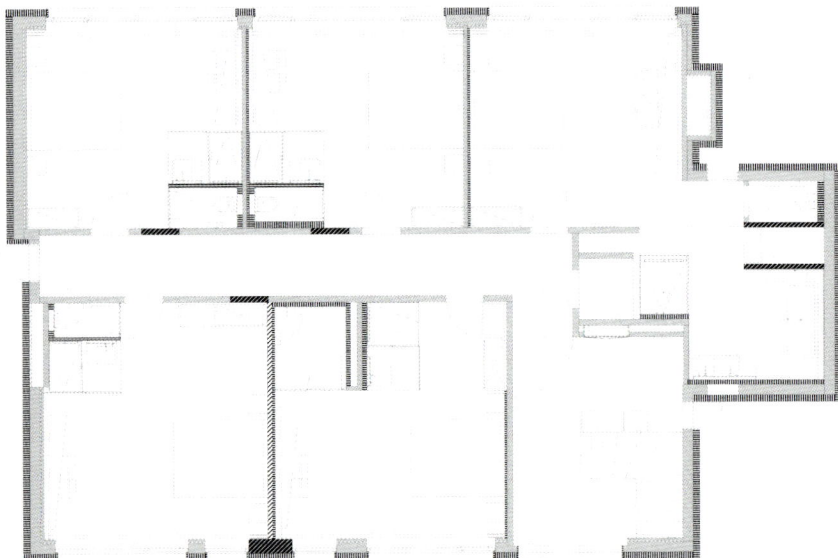

Just about a year ago, the clients decided to renovate the Rössli Hotel in Bad Ragaz. The small but elegantly designed hotel, where the hosts, Doris and Ueli Kellenberger offer seventeen above-average-sized rooms and one suite, has been open since the end of April.

A comfortable, "less is more" atmosphere, with high-quality craftsmanship, was the designers' vision for each room. Space is a luxury which the Rössli can provide! In addition to the large mirrors, focal points in the rooms are the coloured glass walls used between the bedrooms and bathrooms, the exclusive beds and the soaped Douglas firs (spruce) as a floor covering.

大约一年前，位于巴特拉加茨的罗思丽酒店决定进行翻修。这家酒店规模不大，设计却很考究。酒店老板是多丽丝和尤利·凯伦伯格。酒店共有17间超大型客房和一间套房，于4月末开始对外营业。

根据"少即是多"的原则，设计师力图在酒店中创造一种舒适的氛围，在每间客房中展现他们高品质的设计技巧。罗思丽酒店打造了非常奢华的空间。除了巨大的镜子以外，房间中的焦点就是卧室和浴室之间采用的彩色玻璃墙、独享的大床，地面上还铺了道格拉斯枞树（针枞）地板。

2

3

1. Less is more
2. The mirror visually enlarged the space
3. A close-up view of the bedside
4. The windowsill and desk are integrated into one
5. The colour palette as well as the decoration is simple and clear
6. The washroom

1. "少即是多"
2. 镜面扩大室内视觉空间
3. 床头细部
4. 利用窗台设计书桌
5. 色彩和装饰一样简单纯净
6. 卫生间

Hotel Skt Petri
皮特尼酒店

Copenhagen, Denmark
丹麦 哥本哈根

Situated on a beautiful street in Copenhagen's quaint, trendy Latin Quarter, the Hotel Skt Petri is a fine example of superior minimalist Scandinavian design – with a warm welcoming glow. It's a renovated 1930s department store named after the famous church nearby. There are altogether 268 rooms, including 27 suites.

Guestrooms are the most important part for the design of a hotel. In Hotel Skt Petri, every guestroom is to bring guests the most pleasant experiences. There are French windows or outdoor terraces for the guestrooms, so that the surrounding natural scenery is available. The guestrooms are comfortable and clear, where for every detail, function and aesthetics are both taken into consideration.

皮特尼酒店坐落在哥本哈根古怪时髦的拉丁区一条美丽的街道上，是简约的斯堪的纳维亚设计风格的杰出代表，带着一种暖意吸引客人。该酒店原来是一家因附近教堂得名的百货商店，在20世纪30年代重新翻修。酒店共有268间客房，包括27间套房。

客房是酒店设计最重要的环节之一。在皮特尼酒店，每间客房都可以带给客人最舒心的享受。客房有落地窗或者室外平台，让酒店四周的景色尽收眼底。客房内不仅温馨自然，而且每一个细节的设计都是实用与美感结合考虑的成果。

1. The spacious outdoor terrace outside the guestroom
2. The blue-palette guestroom feels tranquil
3. The yellow-palette guestroom feels warm

1.客房有开阔的室外平台
2.蓝色调的客房沉静
3.黄色调的客房温暖

Hotel Suites
套房酒店

Atlantic City, USA
美国 大西洋城

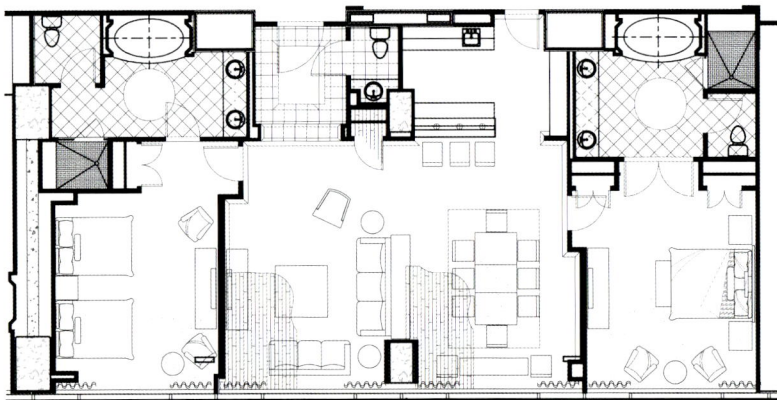

Two types of room styles were designed to attract both a camped customer as well as a retail clientele. Wood floors and hand-tufted carpets created the foundation for the purple-and-red colour schemes.

The End Suites were designed entirely with curved walls that thrust the eyes towards the amazing views of the Back Bay. Custom headboards featured fur and cowhides. The New Orleans Suites were steeped in the tradition of the French-inspired city. The urban sophistication was felt in the core base materials of the rooms' colour palette, but was accented by the celebration of the colours of Mardi Gras. Burgundy, purple and gold were used in various double-colour schemes that offered more surprise and variation than the norm.

套房酒店里设计了两种风格的客房，吸引着临时入住的客人和老主顾。木地板和手工缝制的地毯打造出房间内紫色和红色的基调。

"终极套房"全部采用曲壁，将视线带到外面风景秀丽的"后湾"。特制的床头板上装饰着软毛和牛皮。新奥尔良套房呈现法国传统的都市风情。从套房内材料的颜色基调上能够体验到城市的纷繁复杂，但是突出的是四旬斋的庆典活动的喜庆色彩。深红色、紫色和金色以各种方式两两组合，为我们带来意外的惊喜和变化。

1. The red partitions are the key element in the Chinese-style guestroom
2. The traditional Chinese-style guestroom is quite inviting
3. Living room in the suite
4. Burgundy, purple and gold were used in various double-colour schemes that offered more surprise and variation than the norm
5. The luxurious living room
6. Washroom in the guestroom
7. View of the living room from the bedroom

1.中式风格的客房，红色隔断成为空间亮点
2.带有传统韵味的设计风格让人感到亲切
3.套房起居室
4.深红色、紫色和金色以各种方式两两组合，为我们带来意外的惊喜和变化
5.豪华的起居室
6.客房卫生间
7 从卧室区看起居区

Hyatt Key West Resort & Spa
基韦斯特海厄特度假村与水疗中心

Key West, Florida, USA
美国 弗罗里达州 基韦斯特

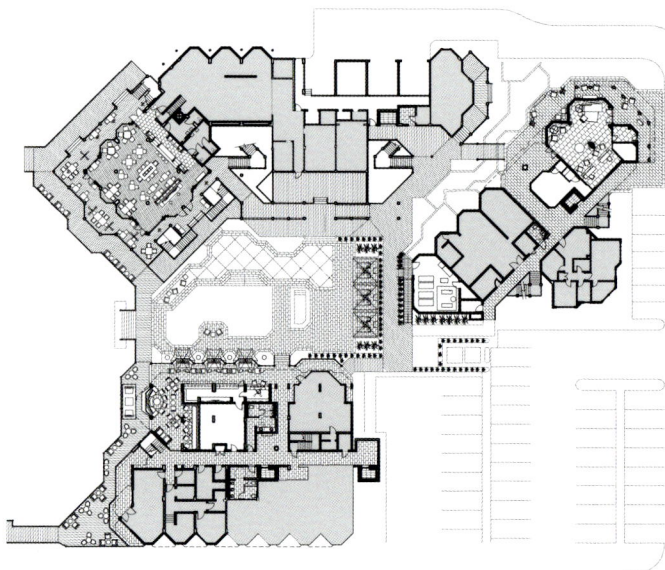

This newly-renovated Key West resort is taking the Key West experience to a new level. As you walk into the guestrooms, you immediately sense the cooling effect of the white-tiled floors. The use of the light-coloured bamboo along with other natural finishes provides the backdrop for the colourful accent fabrics and artwork found in the rooms.

The bathrooms have a spa-like quality with light finishes and unique plumbing fittings. Since the bath opens directly to the room, it acts as an extension of the room with its aqua glass mosaic accent wall and open layout. During the day natural light fills the bath and at night the accent lighted glass mosaic wall adds a dramatic feature to the room.

这家新近翻修的基韦斯特度假村将人们在基韦斯特的体验提升到一个新高度。走进客房，你会立即感受到白色瓷砖地面带来的冷峻感。这里采用了竹子等浅色的自然装饰，为房间内多彩的织物装饰和艺术品提供了背景。

浴室有着水疗室一般的品质，浅色的表面装修和管道的设置十分独特。因为浴室直接连接着客房，所以本身就是客房延伸的一部分。墙面上镶的是浅绿色的马赛克，采用开放式布局。白天阳光会洒满浴室，而夜晚，马赛克墙面在灯光的照明下会为客房增添有趣的一景。

1. The best place for a summer vacation
2. The brilliant guestroom
3. Bathroom in the guestroom

1.尽享阳光假期
2.客房风格如夏花烂漫
3.客房浴室

Hyatt Regency Düsseldorf
杜塞尔多夫海厄特酒店

Düsseldorf, Germany
德国 杜塞尔多夫

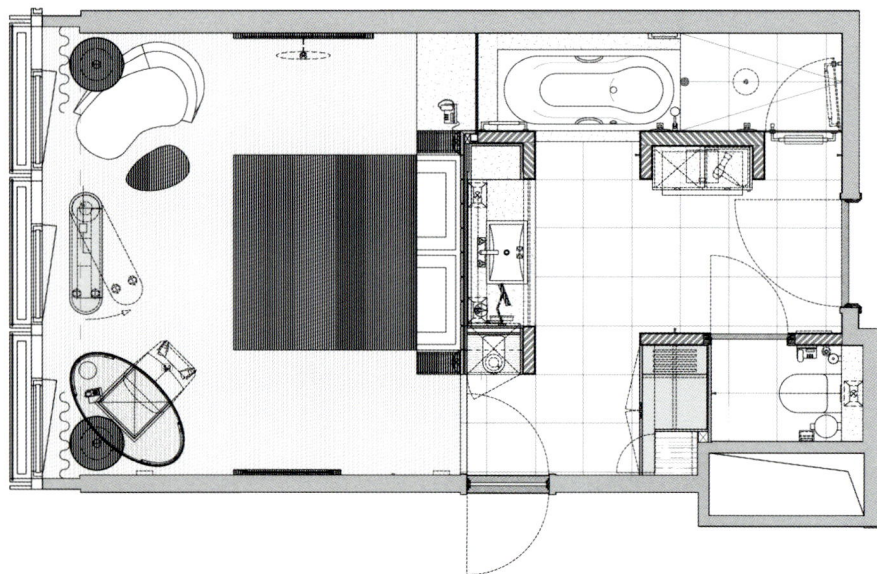

Contemporary and classic in design, this newly-built hotel is located in the trendy Düsseldorf MedienHafen and is home to 303 rooms and suites with contemporary yet comfortable design.

Guestrooms with different styles are available, providing a full range of facilities and services. The comfortable sofas would erase guests' fatigue immediately. The post-modernism paintings on the wall offer a glimpse of the contemporary metropolitan trends. The pendant lamps, the lamps on the bedside, the floor lamps and the wall lamps in the bathrooms are carefully selected, providing guests with the most convenient environment, and more importantly, completing a cosy and intimate atmosphere through lighting.

这座新落成的酒店建筑坐落在杜塞尔多夫的流行商业街上，在设计上结合了现代与古典风格。酒店内有303间客房和套房，设计既现代又舒适。

客房为顾客提供周到的服务，客房内设施齐全，并且有多种风格的客房供选择。舒适的沙发，让客人瞬间忘记疲惫。看见墙上挂着的后现代主义的插画，又可以体会现代都市的时尚气息。屋顶的吊灯，床头的台灯，立式地灯和浴室的壁灯，设计师精心挑选了一系列灯具，一方面营造了舒适的光环境，另一方面也最大程度上为顾客提供方便。

1. The painting on the wall brings a contemporary artistic air
2. Detail of the washroom
3. Lighting design in the bathroom
4. A glimpse of the guestroom
5. The daily commodities prepared for guests also act as decoration for the space

1.墙上插画富有时尚气息
2.卫生间细部
3.浴室内灯光设计
4.客房一角
5.为顾客准备的生活用品也是房间里的装饰品

5

I-Point Hotel

爱点酒店

Bologna, Italy
意大利 博洛尼亚

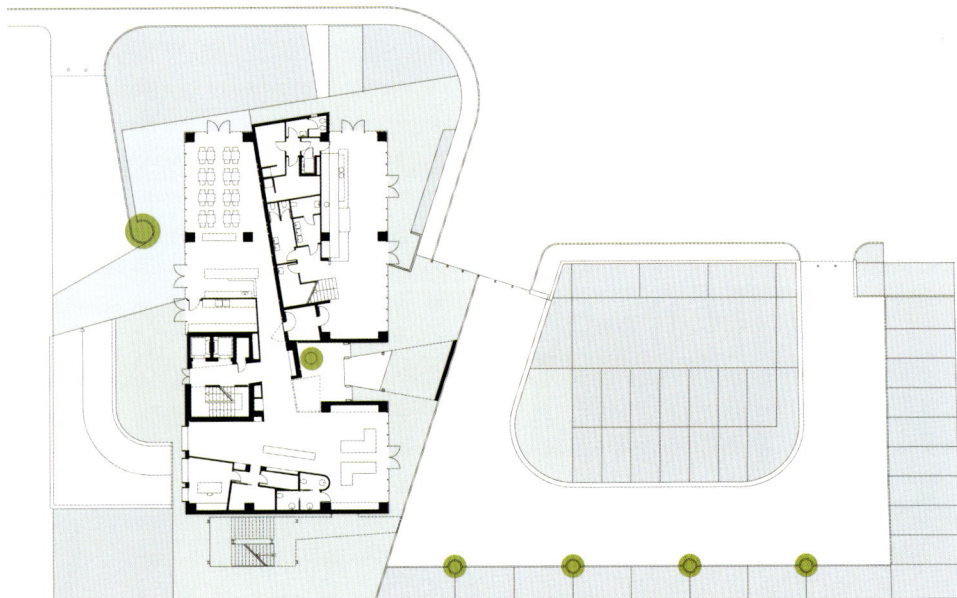

I-point Hotel has been designed for business people, very dynamic, and always on the move, that normally will spend a very short time, normally a couple of nights per stay. Hence the choice of orange and red colour as symbol of energy, is combined with wood finishing, to deliver a different and new, but warm and delightful atmosphere.

Studio SABL re-conceived the original business centre style, delivering a hotel to be experienced down to every single detail: cylindrical-shaped showers, enlightened balcony furnishings, brightly-decorated walls in a balanced contrast with clean relaxation and emotions mixed together. A totally new design concept characterises all and each room, outlining washroom and wardrobe space through transparent walls with an open space effect. This innovative solution allows natural daylight to reach every space within the room, down to the washroom.

爱点酒店专为那些短期居住的商业人士打造，动感十足而又富于变化。橙、红两色使得房间内活力十足，木质装饰则营造出完全不同的温馨恬淡之感。

SABL工作室对酒店原有的商业中心风格进行重新构思，精细到每一个细节：圆柱形的淋浴、阳台上精致的装饰品、亮色装扮的墙壁与房间的淡雅氛围形成和谐对比。全新的设计理念贯穿每一间客房。透明的隔断墙使得浴室和衣橱区增添开阔感，同时使得阳光照射到房间的每一个角落。

2

3

5

1. Double room
2. The terrace outside the guestroom
3. Double room
4. Spectacular view from the guestroom
5. Bathroom in the guestroom

1.双人间
2.客房外的露台
3.双人间
4.客房有开阔的视野
5.客房浴室

KLAUS K

克劳斯K酒店

Helsinki, Finland
芬兰 赫尔辛基

The Klaus Kurki Hotel, a landmark for many years, has been transformed into the Klaus K. Inspired by the emotional contrasts of Finland's national epic, its nature and drama, the hotel bears the stamp of Finland's finest architectural and literary traditions.

Each of the 137 guestrooms is given a theme illustrating the Kalevala's primary emotional elements: desire, passion, mystery and envy. The Klaus K aspires to go further and "take the hotel out of the hotels": creating an ultra-designed lifestyle experience where contrasts abound, such as the Renaissance-inspired space of the Rake Sali ballroom and the playfully-designed theme rooms and suites – the hotel delivers a luxurious experience of tradition and cutting-edge Nordic modernity.

该酒店原名克劳斯·克尔奇酒店，多年来一直是这一区的地标性建筑，现更名为克劳斯K酒店。其设计灵感源于芬兰的历史、自然以及文学之间的对比，彰显芬兰建筑及文学的优异传统。

酒店共包括137间客房，每一间都具有特定的主题，借以诠释不同的情感元素——欲望、激情、神秘以及羡慕。设计师运用现代风格和传统样式的对比，营造了独特的生活体验——文艺复兴时期的舞厅同趣味十足的主题套房格外吸引眼球。

2

3

1. The double room
2. The pattern of the TV wall is eye-catching
3. The TV set is smartly hid in the wall
4. The full-length dressing mirror
5. Local features and traditional styles are integrated
into the guestroom design

1.双人间
2.电视墙的花纹引人注目
3.电视巧妙地藏在墙壁里
4.巧妙设计的穿衣镜
5.带有地域特色和传统韵味的客房设计

Kruisheren Hotel

克鲁舍伦酒店

Maastricht, The Netherlands
荷兰 马斯特里赫特

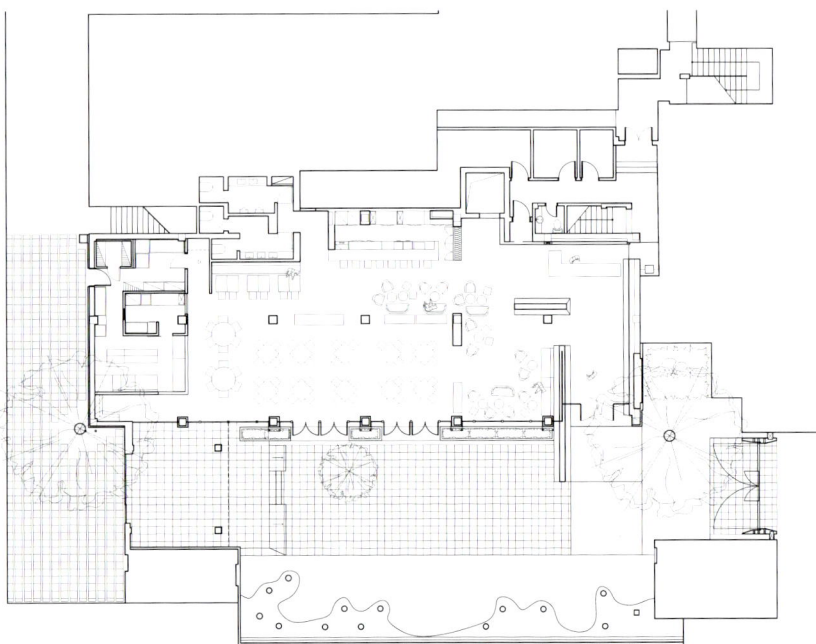

Located in the picturesque Kommelplein Square in central Maastricht, the Kruisheren Hotel is a remarkable conversion of a Gothic church and monastery, a tour de force synthesising the original 15th-century architecture and dressed-down modernism.

The sixty guestrooms are designed with different styles. However, in such a historic building, nostalgia would be an eternal theme. Large black-and-white pictures spread on the walls, demonstrating the cultural history of the building and the city to guests. A bright colour palette is combined, integrating a metropolitan air with the historic building. The orange chairs and the brick-red wall represent the encounter of tradition and fashion.

克鲁舍伦酒店位于风景如画的马斯特里赫特市科梅尔普林广场。酒店从一座哥特式教堂和修道院改造而来，造型十分引人注目，融合了15世纪原始的建筑风格和时尚的现代主义风格。

60间客房风格各异，但是在这样一座历史悠久的老建筑中，"怀旧"成了永恒的主题。大幅的黑白照片布满整个墙壁，让客人感受这座建筑乃至于这座城市的历史文化。与此相结合的是属于现代都市的鲜亮色调，橙色的座椅，砖红色的墙壁，传统与时尚完美结合在一起。

Lánchíd 19

链桥19号设计酒店

Budapest, Hungary
匈牙利 布达佩斯

Named after Budapest's famed "Chain Bridge" spanning the Danube and situated near both it and the Buda Royal Castle, Lánchíd 19 has become a contemporary architectural landmark that attracts a new kind of cosmopolitan crowd.

Lánchíd 19 is a beacon of innovation while still paying homage to its historical settings – a perfect point of departure for discovering Budapest's wonders. Various styles are created for different guestrooms, and correspondingly pieces of different style furniture are selected. In some guestrooms, there are especially-designed wall paintings. The team of Hungarian architects also took every opportunity to exploit the hotel's potential for fantastic views over the city, which can be enjoyed from the suites' terraces or even from the deep bathtub.

链桥19号设计酒店毗邻横跨多瑙河的著名"链桥"（因而得名）与皇家城堡，现已成为这一地区的当代建筑里程碑，吸引着来自世界各地的人群。

酒店在注重营造现代风格的同时也充分地利用了本地的历史特色，可堪称是"寻找布达佩斯特色"的旅程出发点。设计师为不同的客房设计了不同的风格，为每个客房的不同风格精心挑选了不同的家具，一些客房中还有专门设计的墙画。客人在套房的露台以及高大的浴缸内都可以欣赏到布达佩斯城的美景，这是建筑师设计过程中别出心裁的创意。

1. The guestroom brings the warmth of an old film
2. Detail of the wash room
3. Well-equipped guestroom
4. The guestroom offers spectacular views
5. Washroom
6. The green decoration is the continuation of outdoor natural scenery

1.客房老电影一般温暖
2.卫生间细部设计
3.客房设施齐全
4.客房有开阔的视野
5.卫生间
6.绿色装饰是户外自然环境的延伸

Metropole Luxury Boutique Hotel

都市奢华精品酒店

Cape Town, South Africa
南非 开普敦

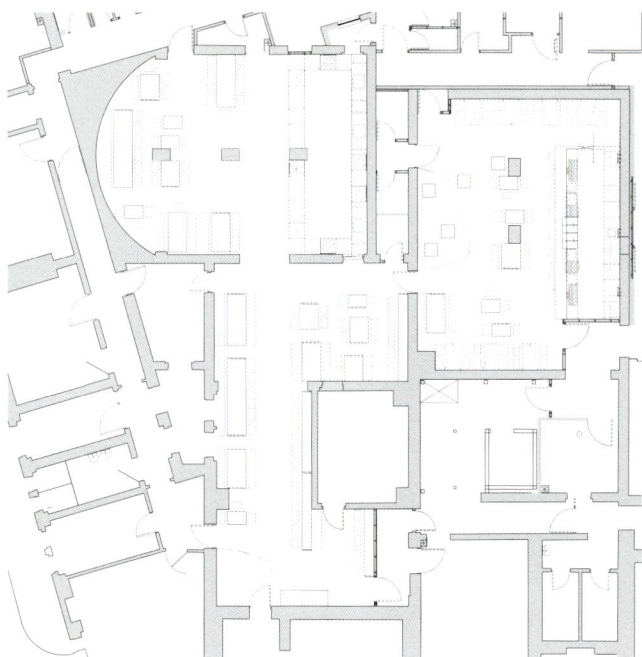

The twenty-six guestrooms provide a contrast of quiet, textured luxury brought up to modern speed by du Plessis's interior concept, which does away with heavy carpets and ageing wallpaper to come up with a soothing minimalism. Nonetheless, the original generous floor plan abides, with its large rooms and wide corridors providing an old-fashioned, spacious luxury – one that has been infused with contemporary elegance and an edgy choice of finishes and flavoured with a vibrant style that is indigenous to the land.

The guestrooms are elegant and graceful. The carpet with simple yet delicate patterns corresponds with the wall with faint patterns in the same neutral colour palette, which continues to the beddings.

酒店共有26间客房。设计师弗朗索瓦·普莱希斯在室内设计中，让静谧的、富于质感的奢华和现代主义风格形成鲜明对比，采用厚重的地毯和古色古香的壁纸，打造了一种令人备觉慰藉的极简主义风格。同时，保留了原有的通透布局，宽敞的房间和宽阔的走廊带来古式的空间奢华感。这种奢华感已经融入了简约的现代感之中，大胆的装饰材料的选择营造出生机勃勃的氛围，这正是开普敦特有的风格。

客房中处处透着优雅，地毯的花纹简单而精致，同色系的墙面有隐约可见的纹理，床上用品也同样是柔和的中性色。

1. The elegant guestroom
2. The painting on the wall of the living area produces an artistic air
3. The double room
4. The wallpaper, carpet, wall painting and furniture are harmoniously combined
5. The bathroom
6. The bathroom

1.优雅的客房
2.起居室墙上的油画增添高雅气息
3.双人间
4.壁纸、地毯、墙画、家具和谐统一
5.浴室
6.浴室

Mission Hills Resort

深圳骏豪酒店

Shenzhen, China
中国 深圳

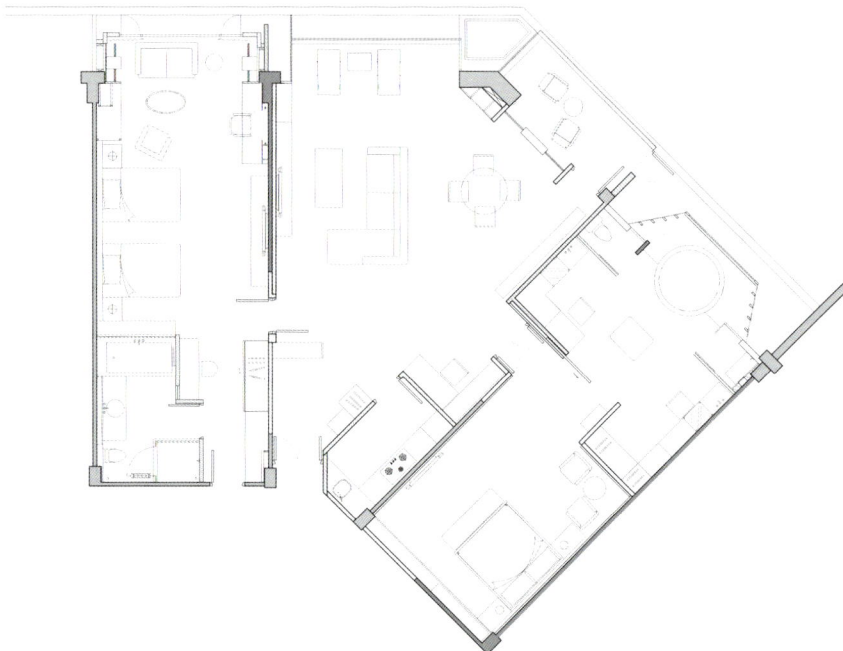

Enjoying the panoramic view of the beautiful golf course, the design takes a subtle yet sophisticated way driven by a "home away from home" concept. Instead of making the design stand out, the designer intended to let is disappear: be it the Premier Room, Deluxe Room and Presidential Suite, everything comes so naturally, so that the guests would feel comfortable and inviting once they enter the space.

In order to create a cosy and homey feeling, the design emphasises greatly on proportion between various compartments and furniture. By taking advantage of natural light which fills up the room, together with the soothing materials and all the delicate details, the designer succeeds in conjuring up a sense of balance, harmony and superiority.

这里有着美丽的高尔夫球场全景，所以酒店的设计在此基础之上巧妙地利用了酒店作为"家外之家"的概念。设计师不让设计超越酒店而存在，而是让设计消失：不管是一流套房、豪华套房还是总统套房，都设计得自然而然，所以客人一进入空间就会感到舒适宜人。

为了打造家一般的舒适感，设计师特别强调了分隔间和家具之间的比例。阳光洒满房间，再加上舒适宜人的材料和各种考究的细节，设计师成功创造出一种平衡、和谐与优越感。

1. The suite enjoys a picturesque view outside the window
2. The suite
3. Detail of the headboard
4. The conference room
5. The study in the suite
6. The luxurious bathroom
7. The private SPA
8. The living area in the suite
9. The dining area in the suite

1.窗外风景如画
2.套房
3.床头装饰细部
4.套房会议室
5.套房的书房
6.豪华浴室
7.私人SPA
8.套房起居室
9.套房餐厅

Mosaic Hotel

马赛克酒店

Delhi, India
印度 德里

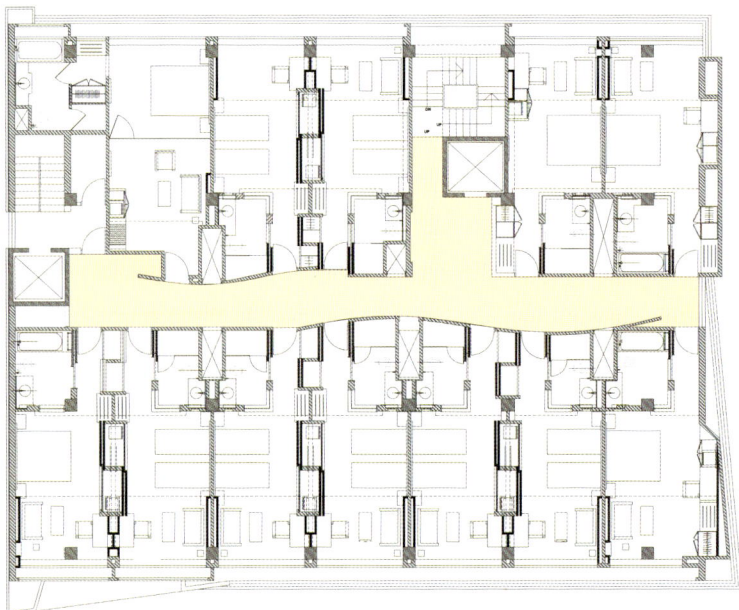

The design of the hotel completely corroborates the name. This is a hotel with a mosaic of experiences, a mosaic of colours, of textures, of lighting, of compositions, of forms, of spaces each with a unique identity and yet integrated together holistically.

To create a much larger feel within the existing small rooms, bathrooms have full-height glass corners open up the room from the vestibule itself. Sliding doors for wardrobes, toilet and refrigerator and cantilevered bedside tables and study table further enhance the spatial feel. Complete edge-to-edge and top-to-bottom glass on the outer side further open up the rooms without a visual barrier from the interior to the exterior. Warm colour tones and textures used in a predominantly white room also allude to the spatial character of the room.

酒店的设计风格完全符合其名字——色彩、灯饰、形状等等全部采用马赛克样式。

客房内，由玻璃结构围和而成的浴室、带有拉门的衣柜、卫生间、冰箱、悬臂式的床头柜以及写字台在视觉上增添了空间的开阔感，而落地玻璃窗的设置以及装饰的温暖色调和质感更是起到了同样的作用。

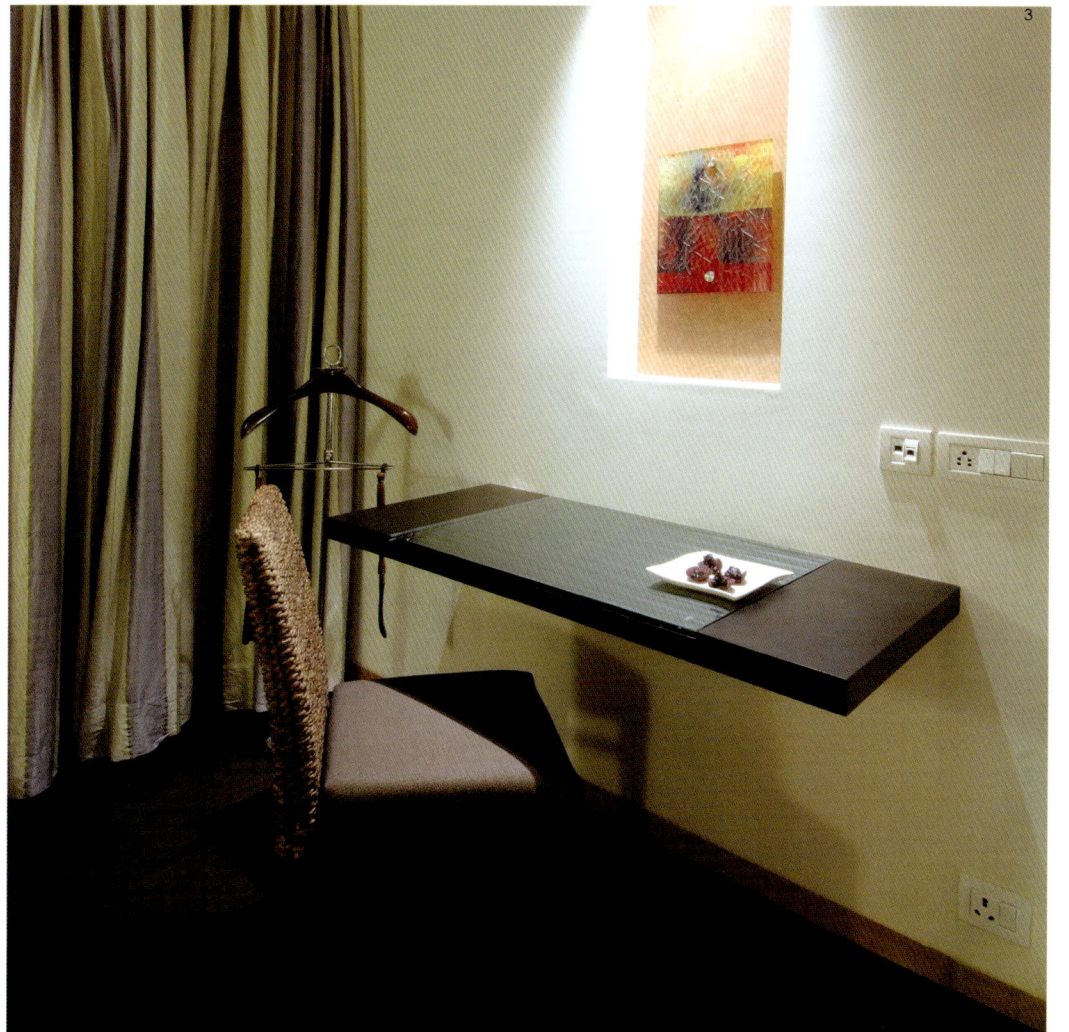

1. The fresh green lighting
2. View of the guestroom from the bathroom
3. The "desk" embedded on the wall is convenient yet space-saving

1.清新的绿色灯光
2.从卫生间看客房
3.书桌方便且不占用太多空间

Palazzo Barbarigo sul Canal Grande
大运河酒店

Venice, Italy
意大利 威尼斯

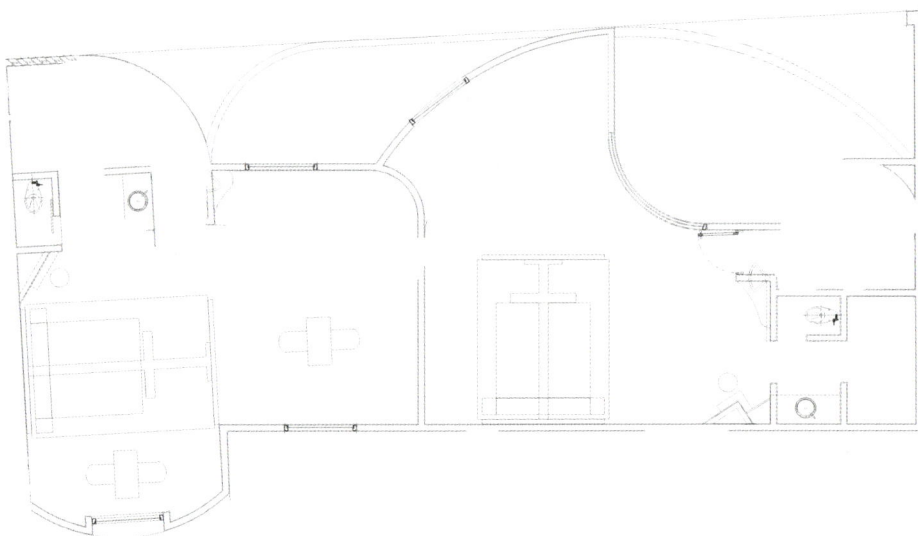

The hotel is located next to the Canal Grande, not far from Rialto Bridge and near the well-known Palazzo Pisani-Moretta, and just few minutes' walk from Piazzale Roma and the railway station. An exclusive haven of comfort recast by designer Alvin Grassi in feminine terms, a blend of past and present spiced with just a hint of the future, and a touch of the mystery that is Venice, the Hotel Palazzo Barbarigo sul Canal Grande is an Art-Deco wonderland that amalgamates both the emotion and playfulness of the Venetian style.

Fine textiles and damask fabrics inspired by Venetian artist Fortuny, an undulating tone poem of understated dove greys, browns and charcoals, furnishings such as curved-leg chairs by Grassi – all of the design touches in this eighteen-room property channel an era of elegant intimacy.

这家酒店临近大运河、里亚托桥和著名的Pisani-Moretta宫殿，距罗马广场及火车站不远，步行只需几分钟。酒店设计融过去及当代风格于一身，同时又带有些许的未来感和威尼斯的神秘之感，可以说是独一无二的修养胜地，舒适而又充满趣味。

艺术装饰风格在每间客房内尽情洋溢——上好的织品搭配缎子材料；灰色、褐色以及木炭色的装饰别具特色，高雅而不失亲切感。

Quilibra Aguas de Ibiza

伊维萨阿瓜斯酒店

Ibiza, Spain
西班牙 伊维萨

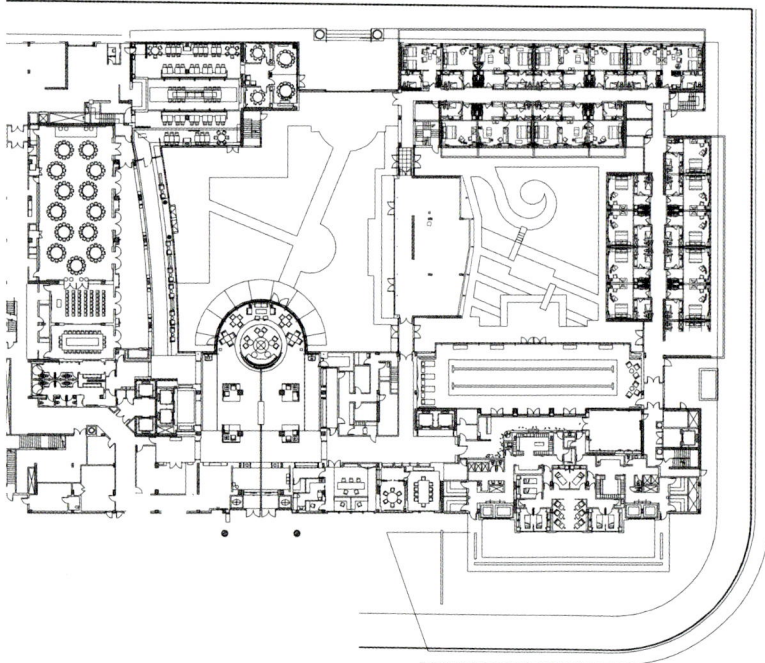

Aguas de Ibiza is the first hotel under the newly launched hotel brand: Quilibra Hotels. This stunning waterside hotel is proudly "anchored" in Ibiza's Santa Eulalia Bay marina. Built by Ibizan architect Juan de los Ríos, designed by Barcelona-based Triade Studio and run by the Torres family of seasoned Spanish hoteliers, Aguas de Ibiza boasts an elegant and modern ambience warmed by a touch of Mediterranean flair and a philosophy of friendly, personalised service.

In all, Aguas de Ibiza offers 112 individually-styled rooms and suites. The interior design by Kim Castells and Jordi Cuenca (Triade Studio) is strongly influenced by the Mediterranean. Dominating colours on the first three floors are blue and white for a crisp and fresh impression whereas in Cloud 9 warm tones like beige and brown are used to create the intimate atmosphere within the luxurious surroundings.

伊维萨阿瓜斯酒店是新建品牌Quilibra酒店集团旗下的第一家酒店，坐落在伊维萨岛圣欧拉利亚海湾，由巴塞罗那的三联工作室操刀设计。酒店因其典雅现代的氛围以及友好的个性化服务而备受青睐。

阿瓜斯酒店共有112间客房和套房，风格各异。室内设计在很大程度上受到地中海风格的影响。最下面三层以蓝、白色调为主，给人时尚清新的感觉；Cloud 9套房内则采用米色和褐色装饰，在奢华的背景中营造亲切感。

1. The white curtains made the double bed romantic
2. The furnishings and ornaments in the guestroom are all carefully chosen
3. The ceiling is particularly designed
4. The suite at night
5. The guestroom enjoys good views of the outside scenery
6. The blue and white colours are typically the Mediterranean style

1.白色纱帘让双人床梦幻浪漫
2.客房里的陈设和装饰都是设计师精心挑选的
3.天花板的设计非常别致
4.夜色中的套房
5.客房可以欣赏到室外美景
6.地中海风格的蓝白色调

Renaissance Las Vegas

拉斯维加斯文艺复兴酒店

Las Vegas, Nevada, USA

美国 内华达州 拉斯维加斯

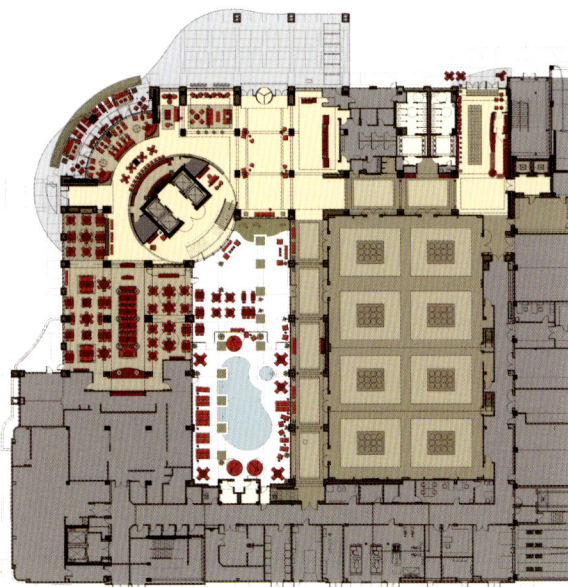

The interior design of the hotel reflects the desire to create a unique hotel for Las Vegas, one that takes the best of Old Las Vegas, and combines it with an environment designed to meet the needs of travellers, both business & personal.

The guestrooms are a little bit nostalgic. Dark green is quite soothing; you would feel as if you could breathe the air in a forest. The patterns on the bedding, carpet and curtain enhance the feeling. The orange sofa and rotating chair remind you of old films, and the small space just feels as warm and comfortable as your home. In the suite, armchairs are set around a coffee table. Family members could sit here, chat, or watch TV, enjoying the lovely time together.

酒店的室内设计反映了"打造拉斯维加斯独一无二的酒店"的愿望，这家酒店要能够展现出古老的拉斯维加斯最具魅力的一面，并满足旅行者的各种需求，包括商务旅行和私人旅行。

客房的设计有淡淡的怀旧风格，墨绿色让人安静，似乎可以呼吸到大森林的气息，床上用品、地毯和窗帘的纹理加强了这种感受。橙黄色的沙发和转椅让人想起老电影，也像自己的家，不大的空间格外温暖。在套房里，扶手椅围着茶几，一家人在这里看电视、聊天，享受生活。

1. The double room
2. Living room in the suite
3. The washroom

1.双人间
2.套房起居室
3.卫生间

Risorgimento Resort

复兴度假村酒店

Lecce, Italy

意大利 拉察

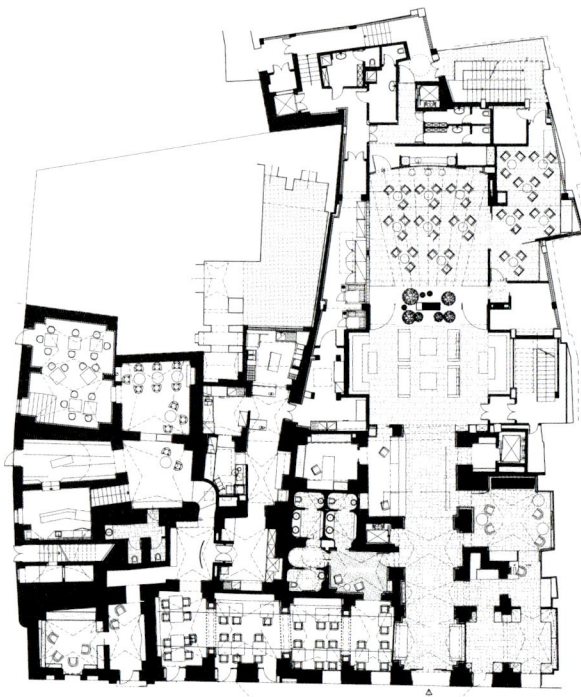

The Risorgimento Resort is a hotel of modern conception encased in a prestigious historic mansion situated in the heart of the magnificence of Baroque Lecce, a few minutes from the bell tower in Piazza Duomo and the Roman Amphitheatre in Piazza Sant'Oronzo. It is unique for the quality of the materials used and the attention lavished on every detail. The hall pays homage to the daily tableau, to a stroll in a street of Lecce Old Town with windows festooned with crepe paper, benches, craftsmen's workshops and bookshops.

On the bedroom walls, large photographic prints on canvas by the photographer Marino Mannarini portray details of Lecce Baroque; the carpets in the bedrooms are patterned in stylised olive trees; every element is intrinsic to the territory and reinterpreted according to contemporary taste.

复兴度假村五星酒店位于老城拉察，距杜奥莫广场的钟塔和罗马圆形剧场不远，步行只需几分钟。天然石材纹理以及历史特色魅力同现代式样家具及灯具结合，营造了非同寻常的效果。

卧室内，墙壁上由摄影师马里诺·曼那里尼创作的巨幅照片彰显了拉察的巴洛克风格；地毯上有时尚的橄榄树图案。总之，每一个细节都凸显地域特色，并根据现代品位进行了重新解读。

2

3

1. Every detail is taken into account
2. The details in the guestroom reveal designers' consideration
3. Outdoor space for leisure
4. The spacious bathroom and the terrece outside
5. The elegant and graceful guestroom

1.精心的设计渗入设计的每个细节
2.客房设计考虑全面周到
3.室外休闲空间
4.开敞的浴室及室外平台
5.客房设计高贵典雅

San Ranieri Hotel

圣拉尼里酒店

Pisa, Italy
意大利 比萨

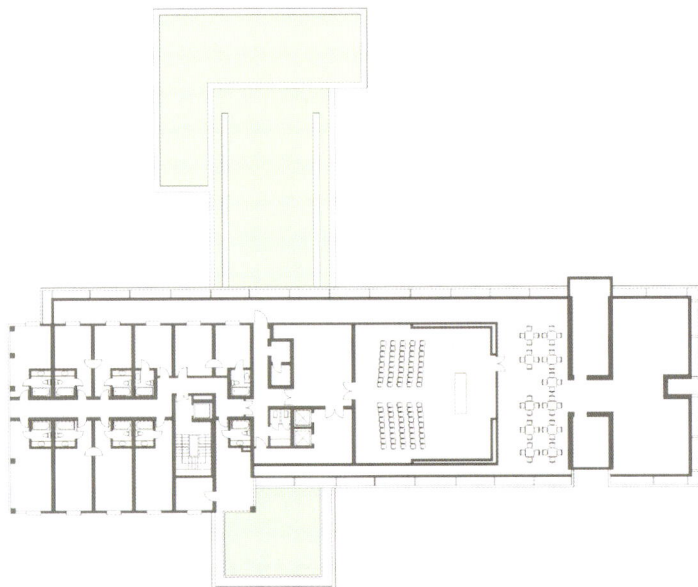

Fruitful collaboration of three Italian Architects who declared through the realisation of this project their purpose of defining the hotel's "new" contemporary tridimensionality, imagined like a resounding oasis for the relaxation of the metropolitan nomadic traveller.

Essentiality, rigour, dynamism, fizz and extraordinary are the messages that this hotel can transfer to the attention of the guests, involving them in a unique and meaningful spatial relationship. Forthright surfaces and charming volumes, interchange with penetrating chromatic lightnings that characterise the visual perimeter, underlined the intent to express contemporary times, beauty and renewed sense of content. The prevailing feeling is that everything can change, that nothing is defined with strict schemes that limit the fantasy.

在这个项目的设计过程中，三位意大利建筑师的通力合作成果颇丰。他们宣称要将这座酒店定义出全新的现代三维效果，可以想象成为大都市的旅行者准备的一片令人瞩目的休闲绿洲。

真髓、严谨、动感、非凡，这就是这家酒店传递给客人的信息，将他们带入一种独特的空间体验中。明朗的表面、迷人的空间以及边缘上的五颜六色的灯光带来的互动感，这就是设计师表达现代感、美以及全新的满足感的方式。最主要的一点就是：每样东西都能变化，没有什么会受到死板的束缚而限制我们的想象。

1. View of the guestroom from the door
2. The living room with French windows
3. The special lighting design creates a mysterious atmosphere
4. The red lighting is quite eye-catching in the black-and-white palette
5. The furniture is designed as an organic whole, which integrates perfectly with the space

1.从入口处看客房
2.落地窗边的起居室
3.灯光从奇妙的角度散射在房间里，空间如同梦幻
4.黑白空间里，红色灯光引人注目
5.一体打造的家具与室内空间完美结合

4

1. The pattern on the carpet is reminiscent of waves of the sea
2. The warm-tone wallpaper enlivens the interior
3. The neutral-palette guestroom
4. The tapestry in the living room demonstrates local characteristics
5. The suite
6. Mirrors are extensively used as adornment

1.地毯的花纹让人想起海上的浪花
2.暖色调的壁纸让室内氛围变得活泼
3.中性色调的房间大方舒适
4.客房起居室，墙上挂毯富有地域特色
5.套房
6.很多地方用到镜面材料做装饰

5

6

1. Every guestroom enjoys different scenes of the sea
2. The Japanese-style interior
3. The guestroom with a private SPA
4. Picturesque scenery is also available in the bathroom

1.每间客房都有不同的海景
2.日式风格的装饰
3.带有私人SPA的客房
4.浴室同样可以欣赏海景

Side

赛德酒店

Hamburg, Germany

德国 汉堡

Side Hotel is located downtown in the city centre, close to shopping areas, the Alster lake, the opera house and local sights.

Special attention is given to materials and light. The headboard is one piece, functioning as bedside tables, lights, wardrobe, minibar and safe. Right down to the telephone and the tooth brush holders, every piece of furniture and decorative element has been selected or developed as part of the concept. All rooms with their soft light and light colours should give you the perfect space for relaxation, enabling you to get up relaxed when you have to leave the Side.

赛德酒店位于汉堡市中心的商业区，离购物区、阿尔斯特湖、歌剧院和当地的许多风景名胜都不远。

特别值得一提的是酒店的材料和照明。床头板是一体的，具备多重功能，包括窗边小桌、床头灯、衣柜、迷你吧台和保险箱。电话和牙刷支托下方的每件陈设和装饰品都是精心设计或选择的，都是设计理念的一部分。所有的客房以其温柔的灯光和淡淡的色调，必将带给你完美的放松空间，让你能够在离开赛德酒店的清晨轻松、舒适地起床。

1. The simple and clean guestroom
2. The open washstand efficiently saves space
3. The living area is laid in a narrow area
4. The innovative furniture in the living room
5. Green plants decorate the lavabo
6. The guestroom with natural light

1.简约素雅的客房
2.开放的洗手台节约了空间
3.狭长区域布置餐厅
4.起居室的创意家具
5.洗漱台上以植物装点
6.阳光下的客房

5

6

The Europe Resort Killarney
基拉尼欧洲度假酒店

Killarney, Ireland
爱尔兰 基拉尼

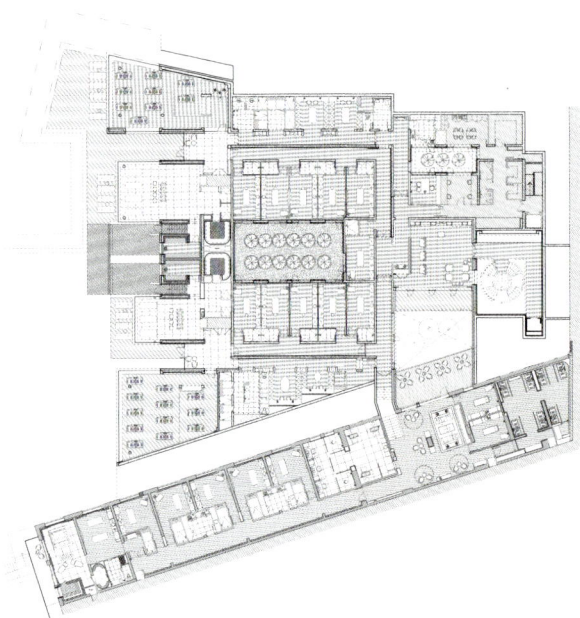

The late-1960s-style, mid-century modern building contained especially spacious rooms that allowed panoramic views of Loch Lein.

In consideration of the local surroundings, the design team brought elements from nature into the property and incorporated them as key interior design elements. Moore commented on the guestrooms, "In creating the palette, we took inspiration from the neutral exterior colours – greys, blues and greens. Wooden tree trunks were collected and fashioned into artwork. All the colours that you see in nature are used as accents in each room, and each room is totally different from the others."

这幢现代建筑始建于20世纪中叶，为60年代后期风格，里面有着特别宽敞的客房，可将罗兰湖的风景一览无余。

考虑到当地的环境，设计团队将自然元素引入了这间酒店，并将它们融合成为主要的室内装饰元素。设计师穆尔对客房的评论是，"在设计色彩时，我们从外界的中性色获得了灵感——各种灰色、蓝色、绿色。我们收集了木制树干，并放进了艺术品当中。你在自然当中见到的所有颜色，都用作了每间客房的色调，而且每间客房都与其他客房截然不同。"

1. Living room in the suite
2. The artistic decorations
3. The aisle in the suite
4. The bedroom enjoys picturesque views

1.套房客厅
2.特色艺术装饰
3.套房走廊
4.窗外景色

The Grand Daddy Hotel

老爹酒店

Cape Town, South Africa
南非 开普敦

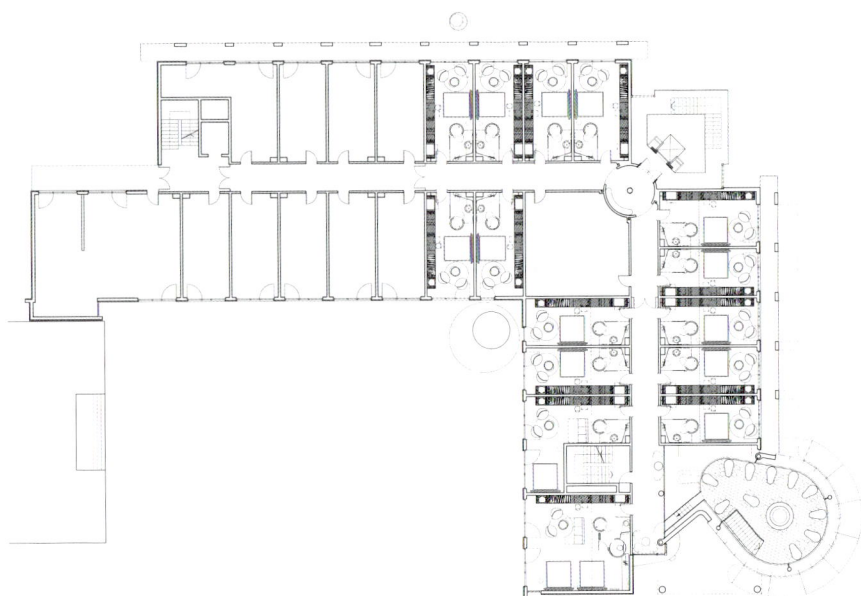

This townhouse was constructed in 1870, acquired a Georgian façade in 1905, and subsequently went through a bewildering, yet brilliant, series of reworkings. In preparation for its current role as Cape Town's The Grand Daddy (formerly the Metropole), François du Plessis made the most of the resulting funky mixture of retro, modern and classic, adding a vivacious touch of African chic.

The twenty-five guestrooms give their visitors the contrast of peaceful luxury brought up to the present day with a soothing, textured minimalism and high-tech appointments. The building might have all the character of maturity and experience, but the hotel that inhabits it boasts an undeniably youthful energy that sparkles exuberantly throughout its spacious chambers and broad corridors.

该建筑于1870年建成，1905年其外观改成了乔治亚建筑风格，后来又经历了一系列的虽令人费解却又辉煌的重修重建，现在又成了开普顿的老爹酒店（旧称京华国际）。为此，设计师弗朗克伊斯·杜·普莱西斯充分把复古、现代和古典这些风格巧妙组合在一起，为建筑物添加了一丝活泼的非洲高雅色彩。

酒店内设25间客房，使参观者体验到把宁静的奢华带到现在生活，既让人放松，同时又运用了高科技。整座大楼既具有所有成熟的特点，同时也有年轻活泼的气质，活力通过宽敞的房间和宽阔的走廊散发出来。

2

3

5

1. A vivacious touch of African chic
2. The dressing table
3. Bathroom in the guestroom
4. Living room in the suite
5. Classic elements are used in the modern design

1.非洲的高雅色彩
2.梳妆台
3.客房浴室
4.套房起居室
5.古典元素运用于现代设计

The Levante Parliament

莱万特国会酒店

Vienna, Austria
奥地利 维也纳

Just behind Austria's parliament building in central Vienna, The Levante Parliament is located around an elegant 400-square-metre courtyard. Guests are also privy to high culture and a little history: in an original Bauhaus building dating from 1908, modern design influences infuse the property, and then there's its function as an art gallery.

In the guestrooms, the dark colour palette feels cool. The embellished orange red is the only bright colour in the interior, bringing a bit of sensibility to the cool atmosphere. The cloth bedding and the furniture both sense dignified, without any additional ornament, highligting the elegance and nobility.

在维也纳的中心，奥地利国会大厦后身，就是这家莱万特国会酒店，坐落在一个占地400平方米的雅致的庭院里。酒店住客在这里还享有高雅文化以及一段历史：这座建筑属于早期的包豪斯建筑，可以追溯到1908年，如今融入了现代设计的元素，赋予了它一种艺术博物馆的功能。

客房中，深色调带来冷艳的感觉。橙红色作为点缀，是唯一的色彩亮点，让冷艳中又多了感性。无论是布艺床品还是家具，都有一种厚重感。没有多余的装饰，更凸显高雅文化的气质。

1. Multiple functions are integrated into one space
2. The wooden desk acts as a partition in the guestroom
3. The washroom
4. The simple yet stylish headboard
5. A close-up view

1.客房空间紧凑
2.客房中的吧台隔断
3.卫生间
4.简约时尚的床头设计
5.细部设计

The Marmara Sisli

马尔马拉西斯里酒店

Istanbul, Turkey
土耳其 伊斯坦布尔

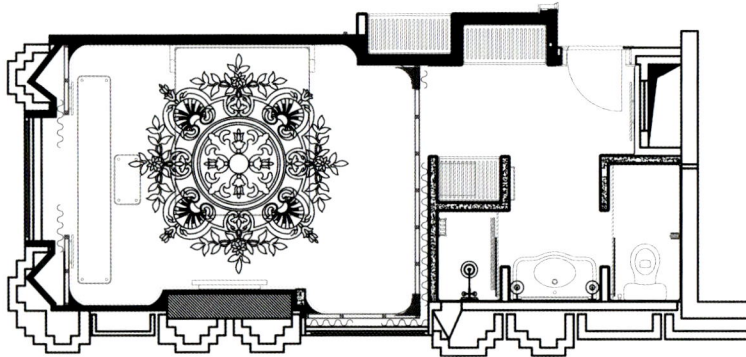

Autoban was commissioned to design the latest home away from home in town. Here you can see Autoban's usual modern contrasts, such as unplastered ceilings with bold patterns.

Following the concept of "less is more", the designers created simple and clear guestrooms. The white beddings and the black lamps on the bedside produce a sharp contrast. Luggage cabinets are hid in the corners, while TV sets are hung on the wall across the bed or stand like billboards in the corners. In the double room, a settee is laid along the wall opposite to the bed, with a perfectly suitable scale for the room. The only decoration in the guestroom is the simple hand-painted patterns on the ceiling.

奥拓班设计公司负责设计了伊斯坦布尔的这家最新酒店。在这里你会看到奥拓班设计公司常用的现代对比手法，比如未抹石膏的天花板（上面有大胆的图案）。

遵循"少即是多"的原则，客房设计简洁大方，雪白的床上用品，黑色的壁灯照在床头的桌板上，行李柜藏在房间的角落里，电视机挂在床对面的墙上，或者像广告牌一样站在角落里。在双人间里，床的对面沿着墙壁是造型简单的长沙发，一切的尺寸似乎是量身打造一样，恰好契合。唯一的装饰是布满天花板的手绘花纹，不过花纹同样简单大方。

1. The pattern on the ceiling is the only decoration in the room, together with the scenery outside the window
2. The simple guestroom enjoys picturesque views
3. The French window beside the bed brings in the beautiful scenery outside
4. The interior space is fully and smartly utilised
5. The washroom

1.天花板上的花纹和窗外景色是室内唯一的装饰
2.虽然房间内简单，但窗外风光美好
3.床边的落地窗可以欣赏窗外景色
4.室内空间被完全利用，十分巧妙
5.洗手间

4. A glimpse of the guestroom, with a peculiar wall painting design
5. The guestroom features a modern metropolitan air
6. The guestroom designed as an organic whole
7. The bright and rich colours bring vigour to the room
8. The green room features a natural air
9. The bathroom finished with mosaic tiles
10. The bathroom with the feeling of countryside

4.客房一角，墙面装饰别出心裁
5.时尚都市客房充满动感
6.一体设计的客房
7.亮丽的色彩充满活力
8.绿色客房春意盎然
9.马赛克装饰浴室
10.田园气息的浴室

6

7

8

The Three Stork Hotel

三鹤酒店

Prague, Czech Republic
捷克共和国 布拉格

The five-star hotel is situated in the historical heart of Prague right below Prague Castle. During the reconstruction all the historical fabric such as the plastering, painted ceilings, floors, valuable baroque roof construction were preserved and restored to their original condition. The hotel is furnished with fitted modern furniture and the seating by Moroso and B&B Italia.

Each room is original and some have glass bathrooms and some are fitted with a special furniture unit containing a bathroom. On the basis of common neutral colours, each room features its own bright colour. Each room distinguishes from the others by the design of the wall, the ceiling, the layout and furnishings.

五星级的三鹤酒店坐落在古老的布拉格中心——布拉格城堡下。在重建阶段，原有的历史特色元素，石膏、漆顶、地板、巴洛克艺术风格屋顶都完全保留下来，并彻底恢复了其本来的面貌。酒店内，家具及座椅由意大利知名的Moroso 和B&B公司制作，现代风格十足。

客房各具特色，有的设有玻璃浴房；有的布置了特色家具。中性色调的底色上，每个房间选择不同的亮色作为主题。墙面和屋顶的设计以及空间布局、家具陈设也强调了每个房间不同的特色。

1. The room in a yellow tone feels warm and comfortable
2. The unique ceiling design in the room
3. The room in a red tone feels spacious and fervent
4. The unique ceiling design, making advantage of the top floor
5. Washroom

1.黄色调的客房温暖舒适
2.客房房顶的设计独具匠心
3.红色调的客房宽敞热烈
4.充分利用顶层特点设计屋顶别具一格
5.卫生间

The Vine Hotel

藤蔓酒店

Funchal, Madeira, Portugal
葡萄牙 玛德拉群岛 丰沙尔

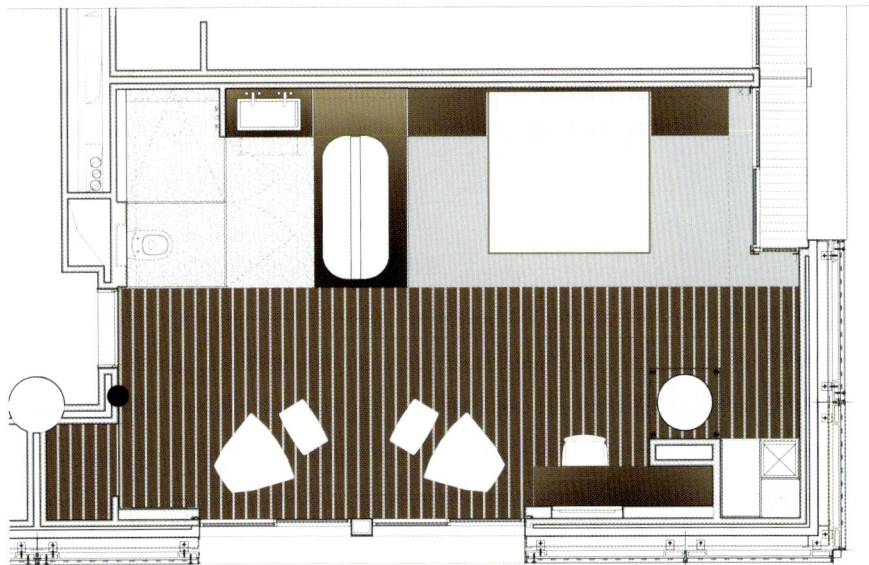

The Vine Hotel was created on the concept as its proper name implies. Each floor with the different colours symbolises the four seasons.

In the guestrooms, glass is the main material, together with other decorative materials, creating a stylish visual experience for guests. The tawny glass partition brings us a "flowing" space. The lighting design here is also marvelous. The lamps hid in the ceiling cast inviting light, reflecting on the glossy surfaces of the furniture. The whole room looks like a giant piece of artwork, giving off mysterious luster. The pure white bed in the centre of the guestroom is highlighted.

藤蔓酒店的设计理念遵循酒店名字的含义。每层楼上采用一种不同的色彩，代表四季。

客房里，玻璃、其他装饰材料作为主要装饰元素，打造客房时尚的视觉感受。茶色玻璃作为隔断，通透而富有变化。灯光设计同样妙不可言，隐藏在天花板上的灯在夜晚散发出奇妙的光，在家具光洁的表面上散射，让整个客房如同一件巨大的艺术品，散发出神秘的光泽。纯白的床在房间正中，被衬托了出来。

1. The tawny glass partition brings a special visual experience
2. The open bathroom
3. The bathroom
4. The guestroom is extremely romantic with the petals scattered
5. The space is simple yet stylish
6. View of the bathroom from the living area

1.茶色玻璃隔断打造特别感受
2.开放的浴室
3.浴室
4.花瓣洒满客房，别样的浪漫
5.简约时尚的设计
6.从起居室看浴室

4

5

The Wit Hotel

风趣酒店

Chicago, USA
美国 芝加哥

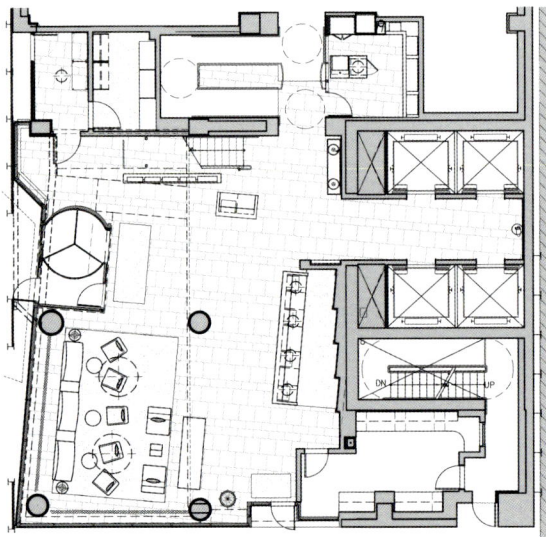

At the heart of the legendary Loop, where the "L" passes just above the office tower doors, the life of a great city – its grit, its glamour, its great and famous wind – inspires an exciting new hotel that lives up to its name, the Wit.

The guestrooms are also designed with wit. The designers strived to spread the theme of wit with the marvelous choices of colours, the novel decorations, etc. In such guestrooms, guests would have the most enjoyable time when they find constant surprises in every corner.

风趣酒店位于芝加哥著名的"Loop"中心，这个"L"正好经过办公楼大门的地方。这座大都市的生活方式——它的坚毅、它的魅力以及它著名的风——都成为设计这座全新的酒店的灵感，这家酒店实至名归，充满了"风趣"。

客房的设计同样充满"风趣"，设计师们充分发挥想象力，无论是奇妙的色彩搭配还是别出心裁的装饰都紧扣"风趣"这一主题。客房布置不落俗套，让顾客在每个角落都可以发现惊喜，享尽愉悦时光。

1. Close-up of the bedside   1.床头细部
2. The furnishings are "witful"   2."风趣"的陈设

T Hotel

T酒店

Cagliari, Italy
意大利 卡利亚里

T Hotel rises out of Cagliari's future cultural centre: the Parco della Musica, a verdant area with fountains and gardens, in which the hotel and the Teatro Lirico hold centre stage, will also contain an open amphitheatre, a contemporary art space and a theatrical stage design laboratory.

T Hotel features a total of 207 rooms of different sizes, which allow for very customised hospitality where harmony and functionality join to offer very striking living solutions. Comfortable king-size beds are also part of the focus on relaxation. There are various types of rooms ranging from the Classic, the Deluxe, Junior Suites and large 55-square-metre Suites in the tower, with breathtaking views of the city and the sea, so that diverse requirements can be fulfilled. The hotel also features "friendly rooms" that are set up so as to cover the requirements of disabled or elderly people.

四星级T酒店位于卡利亚里市未来文化中心内，四周环绕着音乐大厅、绿地花园、圆形剧场、现代艺术馆及舞台设计室。

酒店共包括207间客房，规格不一，能够满足客人的不同需求，提供舒适的环境。超大号的床格外舒适。此外，客房的风格也多种多样。酒店还设有专门供残障人士和老年人使用的无障碍房间，极为人性化。

4

5

1. The well-equipped guestroom
2. The guestroom with a jump-layer
3. Detail in the guestroom
4. The elegant green
5. The elegant and comfortable guestroom
6. The flowers bring a spring air to the room
7. A glimpse of the guestroom
8. Bathroom in the guestroom

1.客房设施齐全
2.跃层客房
3.客房细部设计
4.淡雅绿色
5.客房优雅舒适
6.鲜花让房间春意盎然
7.客房一角
8.客房浴室

Uma Paro
帕罗乌玛酒店

Paro, Kingdom of Bhutan
不丹 帕罗

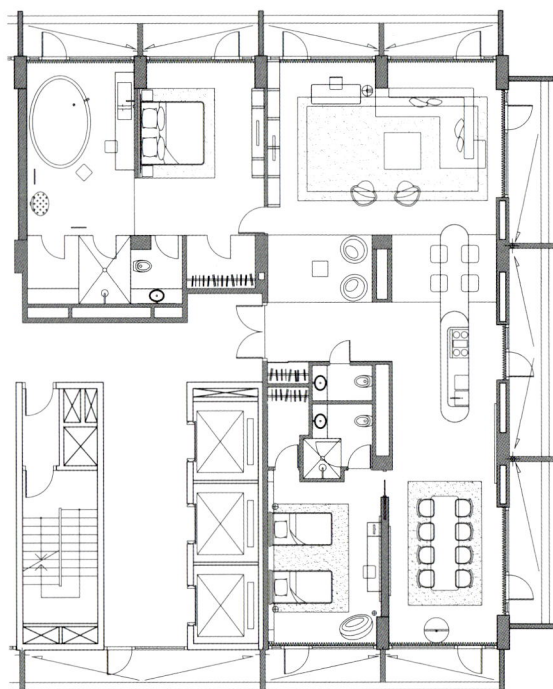

The Bhutanese king's conscious policy of protecting the country from tourism has helped keep the Himalayan nation refreshingly pristine, and the designers of Uma Paro embraced this unspoiled quality. The resort was designed in collaboration with traditional Bhutanese artisans: indigenous detailing adorns interiors; walls are hand-painted by local artists.

The twenty-nine rooms and nine villas provide dazzling views of the Paro Valley below, including the rice paddies. All areas incorporate as much sunlight and visible nature as possible so that guests can easily access the magical mountain kingdom that surrounds them. The nearby town of Paro offers them the chance to experience firsthand the country's thriving Buddhist culture.

不丹国王的政策有意强调保护国家不受旅游的侵害，这使得这个喜马拉雅国家保持着未受影响的原始状态，帕罗乌玛酒店的设计师正是利用了"原始"这一特点。度假村的设计结合了传统的不丹技工：具有当地特色的细部点缀着室内，墙面是当地艺术家手绘的。

29间客房和9栋别墅都能看到帕罗峡谷的美丽景色，包括几片稻田。所有的空间都尽量多地利用阳光和大自然，这样酒店客人就能够很容易地接近围绕着他们的神奇的山脉王国。邻近的帕罗镇让他们有机会去切实体验这个国家繁荣的佛教文化。

5

6

1. The guestroom enjoys broad views
2. The bathroom
3. The windows spread on the whole wall broaden the views
4. Picturesque views are also available in the bathroom
5. The suite
6. The hand-painted wall design is a key element

1.客房视野开阔
2.浴室
3.布满整面墙的窗户扩大了视野
4.浴室同样可以欣赏窗外景色
5.套房
6.手绘墙壁是亮点

UNA Hotel, Bologna

博洛尼亚UNA酒店

Bologna, Italy

意大利 博洛尼亚

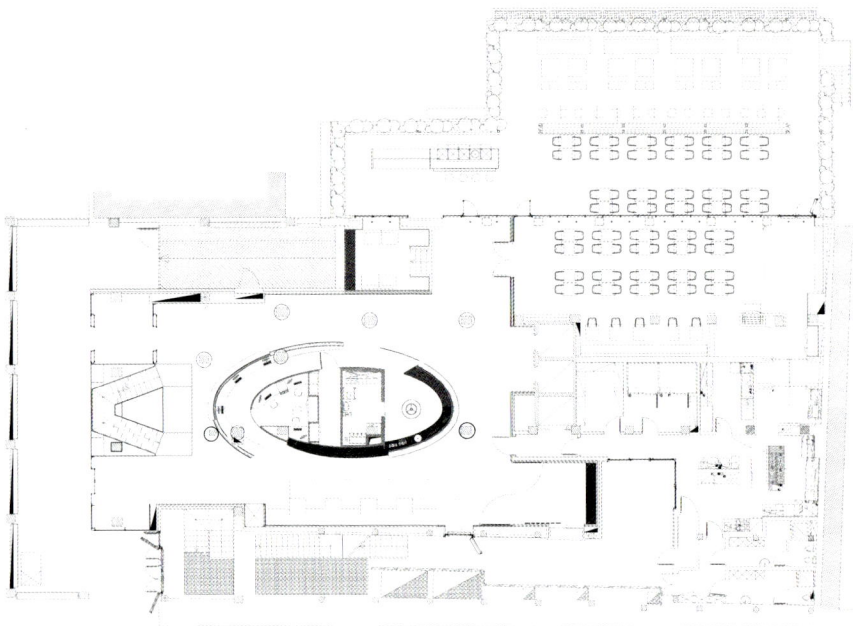

Like any other meeting and exchange venue, Bologna functions as a crossroad of habits, customs, languages and thus also of writings. It therefore seemed to designers to be consistent to adopt writing as the unifying and descriptive element of this new "place" dedicated to tourism. This is why certain languages/writings were chosen to denote the convergence of the various cultures.

The graffiti in the corridor, the bright colour palette, the simple furnishings and the exquisite lighting are combined together to create the modern guestrooms. Meanwhile, texts are also adopted here. From some of these written accounts they chose an excerpt, then a phrase linked to a particular place or landscape, and then, finally, the letters that make up a word to create a graphic device that characterises the hotel's settings and accompanies the contemporary traveller.

同其他文化交流中心一样，博洛尼亚是各种风俗习惯以及语言文字的集散地。设计师正是从这一点上获得灵感，将文字作为主要设计元素，借以传达各种不同的文化。

走廊里的涂鸦、客房中鲜明的色调、简单流畅的造型和精致的灯光设计带来了极富都市感的客房设计。同时，文字元素也运用在客房中，从精选介绍中提取关于地方及风景描述的摘要和短语，然后将字母拼写成单词，组成各种平面图案，构成酒店的特色背景，让那些现代的旅行者耳目一新。

1. The text decoration on the headboard
2. Tables and chairs for dining and conversation
3. Bathroom in the guestroom
4. The dreamy blue
5. The comfortable guestroom

1.床头的文字装饰
2.可供闲谈就餐的桌椅
3.客房浴室
4.蓝色梦幻
5.舒适的客房

UNA Hotel, Naples

那不勒斯UNA酒店

Naples, Italy
意大利 那不勒斯

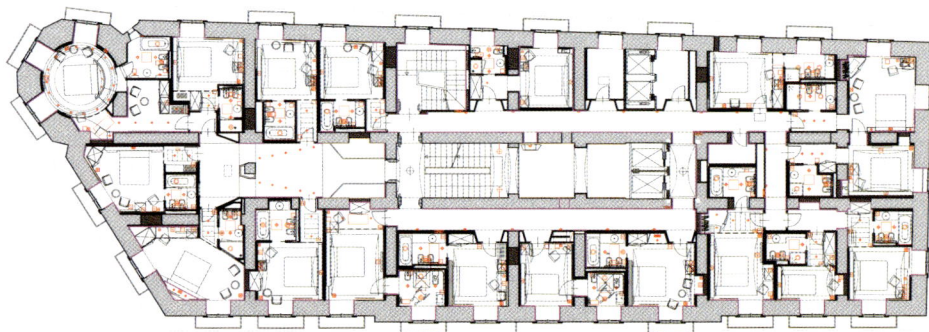

A 19th-century historic building, facing on Garibaldi Square into the city centre, has been restored and transformed into a four-star hotel for a prominent Italian hotel chain. The original building, a seven-storey tuff-stone structure, very long and narrow, has suggested the design of a vertical hall crossed by staircases and lifts dangling in space. The structure hosts ninety rooms, a small convention centre and restaurant with a roof garden overlooking Vesuvius.

The guestrooms are designed to bring the sense of entering a palace. There is no complex classical decoration, but the designers bring out similar effects with modern ways of decoration. In addition, the designers made use of the bizarre spaces in the old building to offer new diversified interior spaces.

那不勒斯UNA酒店隶属于意大利知名酒店连锁集团，由一幢始建于19世纪的建筑改建而来，共包括90间客房、一间小会议室以及带有屋顶花园的餐厅。原建筑呈现窄长形状，共7层。设计师根据这一特点布置空间格局，运用交叉设置的楼梯和电梯连通上下。

客房的设计会带给客人一种步入宫殿的错觉，虽然没有古典设计繁复的装饰，但是设计师巧妙地用现代设计手法带来了类似的效果。老建筑中的异形空间也被设计师巧妙地利用，创造性地设计了富于变化的室内空间。

1. Two broad doors on both sides of the head of the bed
2. Spectacular views of the city outside the window are available on the bed
3. The living room
4. The elegant guestroom
5. Bathroom in the guestroom

1.巧妙利用多门的转角
2.在床上就看得见窗外的城市
3.起居空间
4.客房空间优雅
5.客房内浴室

URBN Hotel

上海雅悦酒店

Shanghai, China
中国 上海

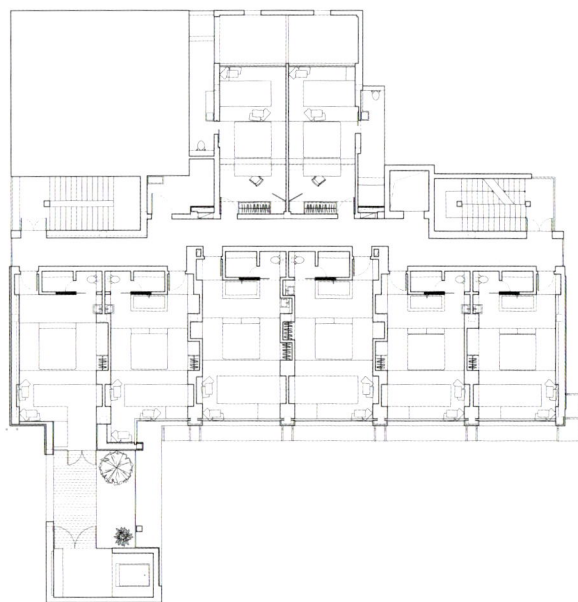

In May 2007, URBN Hotels entered into an agreement with Climate Bridge, an international intermediary offering bespoke solutions for companies and industries to reduce and offset greenhouse gas emissions. The total amount of energy the hotel consumes, including staff commutes, food and beverage delivery, and the energy used by each guest, will be tracked to calculate the carbon footprint.

To be consistent, the lighting in the hotel is particularly designed to be low energy-consuming. Such is true in the guestrooms where efficient lightbulbs are installed. The custom-made furniture in the guestrooms are simple but delicate. The designers made full use of the uneven floor to create different layers in the space. The interior decorations are all in the Chinese style, being simple and noble at the same time.

2007年5月，雅悦酒店与"环保桥"组织（一个帮助各公司和行业减少并抵消温室气体排放现象的国际组织）签订了一份协议。酒店消耗的全部能量，包括员工通勤、食品和饮料递送、每个客人使用的能量，都将会被记录并用来计算碳的排放量。

紧扣酒店的设计主题，酒店内都使用低瓦数照明，客房内同样使用节能灯。客房内定制的室内设施简洁别致，设计师巧妙地利用地平的起伏制造了空间的层次。室内装饰风格都是低调而高贵的中式风格。

Wynn Macau

永利澳门酒店

Macau, China
中国 澳门

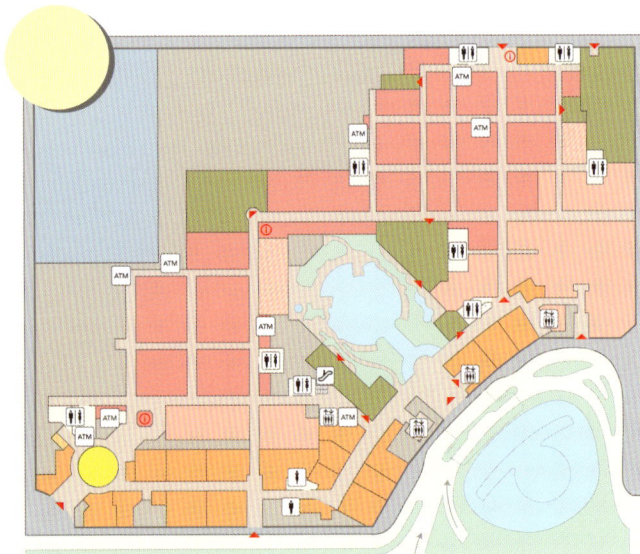

Wynn Macau has 600 rooms and suites. There are 360 deluxe rooms and 100 grand deluxe rooms, with an average of 56 square metres of luxury living space. The resort has 120 one-bedroom suites, with an average of 185 square metres, and 20 two-bedroom suites, with an average of 278 square metres, including lavishly appointed bedroom and entertainment space.

The guestrooms are luxurious, like other spaces in the hotel. The custom-designed carpet is quite eye-catching with special patterns. Carefully-chosen pieces of artwork are presented in the guestrooms and corridors. The wooden furniture is characterised by the delicate carved patterns. The heavy curtains offer a special texture and luster. At night, the scenery of the sea is attractive, and the night view of bustling Macau is really beautiful.

永利澳门酒店共有600间客房和套房，其中360间豪华客房，100间特级豪华客房，每间面积为56平方米。还有120间一室套房（面积为185平方米）及20间两室套房（面积为278平方米），包括豪华的卧室及娱乐空间。

如同整个酒店的设计风格，客房设计也非常奢华。定制的地毯有抢眼的花纹，房间和走廊里都摆满设计师精心挑选的艺术品，木质家具上有精致的雕刻，厚重的窗帘有独特的肌理和光泽。夜晚，窗外有海景，不远处就是澳门繁华的夜景，非常漂亮。

1. The comfortable chairs in the corridor
2. Beautiful night view is available outside the window
3. Bathroom in the guestroom
4. Bathroom in the guestroom
5. Living room in the suite

1.走廊上的休闲座椅
2.夜色美景
3.客房卫生间
4.客房卫生间
5.套房起居室

X2 Kui Buri

奎汶里X2度假村

Petchaburi, Thailand
泰国 碧武里

The general area is underdeveloped and peaceful with the only communities being local fishing villages. The 23 semi-private villas each have their own terrace, garden and most with private pools.

X2 Kui Buri is designed by Duangrit Bunnag. Each semi-private villa is sculptured in mountain rock walls, with expansive glass-panelled doors and generous bedroom and bathroom spaces. Furnishings and fixtures in the villas were chosen to be consistent with the X2 concept for luxury in design and function. The pool villas showcase a private plunge pool ranging from 20 square metres to 30 square metres in size and also a private terrace and garden courtyard.

整个这片区域并没有太复杂的设计，显得很宁静，唯一有的就是当地的一片渔村。23座半私人的别墅每座都有自己的露台和花园，大多数还有私人泳池。

奎汶里X2度假村由巴纳格设计完成。每间半私人的别墅都是嵌在山上的岩石里，有宽敞的玻璃大门，大气的床和浴室空间。别墅内的家具陈设都是特别选择的，风格跟X2度假村的设计理念一致，那就是在设计和功能上都要体现奢华。有些别墅带泳池，仿佛私人瀑布水潭，面积从20平方米到30平方米不等，此外还有私人露台和花园庭院。

1. The guestroom enjoys abundant warm sunshine
2. The open-air bathroom
3. Detail of the headboard
4. The neat stand
5. Bathroom in the guestroom

1.客房外就是温暖的阳光
2.露天浴室
3.床头细部
4.巧妙的置物台
5.客房浴室

Yas Hotel

雅思酒店

Abu Dhabi, UAE
阿拉伯联合酋长国 阿布扎比

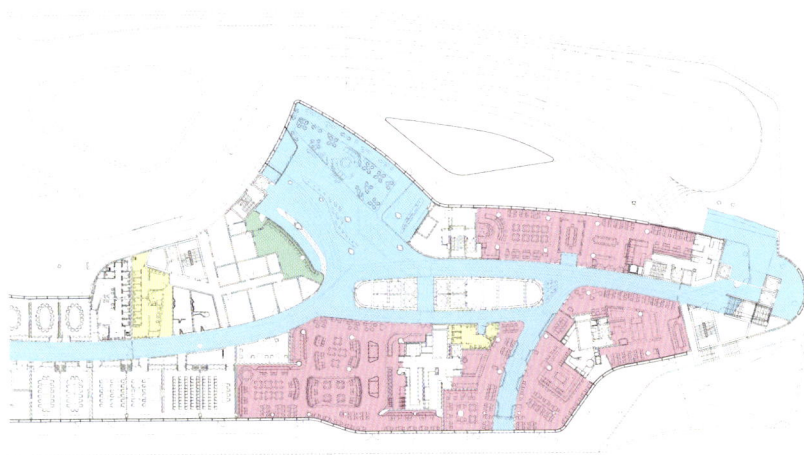

Once inside the room, any division between the sleep zone and the wet zone is further dissolved, to create a single flowing living space, with functionality defined only by finishes.

The materials include leather-textured porcelain tiles or ribbed, Carrara marble tiles with tinted bronze mirrors and inset, back-lit alabaster wash stands. The bath is topped by a heavy slab of pure white quartz, which extends seamlessly into the sleep zone as an amorphous, flowing plinth which clasps the bed and the lounger. The finishes in the sleep and work zones are natural, but refined. Fine joinery in sun-bleached oak is shaped in reference to the form of wind-filled sails of local dhows. Unruly rugs define a "sit zone" and porcelain tiles are pressed with stone aggregate.

一进入客房，睡眠区和洗浴区的分界线似乎消失了，整个是一个单独的流动空间，各个功能区仅用不同的装饰来区分。

材料方面，有皮质肌理的瓷砖和有棱纹的卡拉拉大理石饰面砖，上面带有彩色铜镜和嵌入的背光照明的条纹大理石洗手盆台面。浴室上面是一块白色的纯石英板，一直延伸到睡眠区，成为一个基座，下面是床和躺椅。睡眠区和工作区的装饰自然而又考究。晒白的橡木制成的精致的细木手工艺品造型好像当地一种独桅帆船鼓风的帆。独特的小块地毯界定出起居区，石材上铺着瓷砖。

1. The guestroom is a single "flowing" space
2. Living area in the suite
3. The luxurious guestroom
4. Bedroom in the suite
5. Standard room

1.客房是一个单独的流动空间
2.套房起居室
3.豪华客房
4.套房卧室
5.标准间

Yinchuan Kempinski Hotel

银川凯宾斯基酒店

Ningxia, China
中国 宁夏

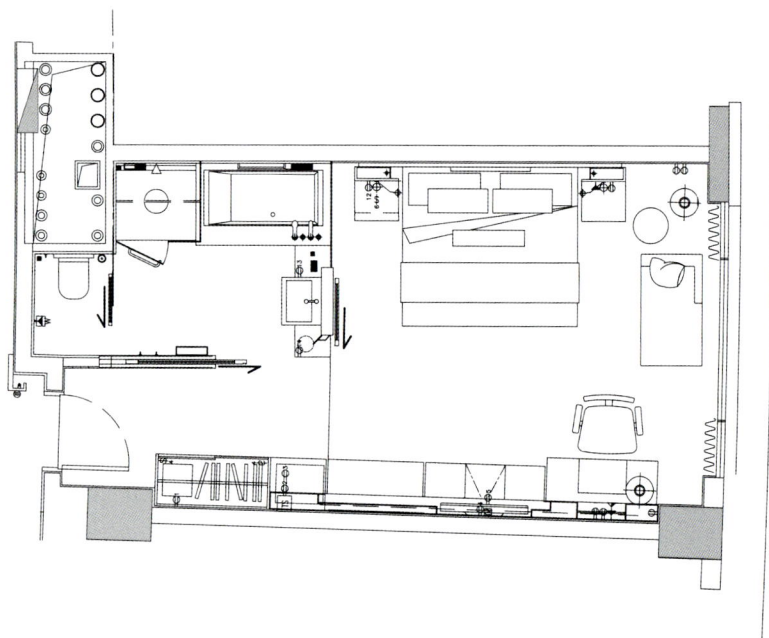

The guestroom is the heart of the hotel experience and here the designers reinforce the Kempinski Brand while providing a comfortable residential environment for the international traveller. The guestroom will have the latest in contemporary luxury with a bathroom that can be completely opened up to the bedroom. Timeless colour schemes of taupe, beige, chocolate and smoky green combined with walnut wood, zebra wood and contemporary furniture are the core of the room palette. The guest corridor is an extension of this tailored sophistication. Here walnut doors with concealed light at the guestroom entrance and a streamlined wall sconce, provoke a sense of sophistication and create a high expectation for the guestroom experience.

客房是酒店的灵魂。在这家酒店中，设计师在为国际旅行者打造了舒适的居住环境的同时，强调凯宾斯基品牌。客房里呈现出最新的现代奢华流行风。浴室能打开，跟卧室相通。经得起时间考验的色调——灰褐色、淡棕色、巧克力色、烟熏绿——结合在一起，搭配胡桃木、斑马木和现代家具，共同构成了客房的核心基调。客房外的走廊延续了这种量身定做的舒适感。胡桃木的客房门，玄关处隐藏起来的灯具，流线型的壁突式烛台，都创造出一种精致的奢华感，让人对客房的入住体验备感期待。

1. The stripe pattern on the carpet corresponds with that on the wooden material
2. The zebra wood is also used in the bathroom
3. The desk is convenient for computer users
4. The wall is fully and smartly explored, with a certain aesthetic feeling
5. Detail of the bedside

1.地毯的条纹与装饰材料的木纹相呼应
2.浴室中同样用到了这种斑马纹的木材
3.客房配备了可放置电脑的书桌
4.墙壁被充分利用，设计巧妙而不乏美感
5.床头细部设计

图书在版编目（CIP）数据

精品酒店空间设计系列. 客房与卫浴 / 谢昕宜编 ; 李婵译.
— 沈阳 : 辽宁科学技术出版社 , 2015.3
　　ISBN 978-7-5381-9078-6

　　Ⅰ .①精… Ⅱ .①谢… ②李… Ⅲ .①饭店－室内装饰设
计－图集 Ⅳ .① TU247.4-64

中国版本图书馆 CIP 数据核字 (2015) 第 028056 号

出版发行：辽宁科学技术出版社
　　　　　（地址：沈阳市和平区十一纬路 29 号　邮编：110003）
印　刷　者：沈阳天择彩色广告印刷股份有限公司
经　销　者：各地新华书店
幅面尺寸：225 mm×295mm
印　　张：34
字　　数：50 千字
出版时间：2015 年 3 月第 1 版
印刷时间：2015 年 3 月第 1 次印刷
责任编辑：常文心
封面设计：何　萍
版式设计：迟　海

书　　号：ISBN 978-7-5381-9078-6
定　　价：280.00 元

联系电话：024—23284360
邮购热线：024—23284502
E-mail:lnkjc@126.com
http://www.lnkj.com.cn
本书网址：www.lnkj.cn/uri.sh/9078

呦呦鹿鸣

燕国公主眼里的霸国

Harmonious Life:
The State of Ba in the Eyes of
a Yan Princess

山西省考古研究所
山西博物院　编
首都博物馆

科学出版社
北京

首都博物馆 书库

丁种 第贰拾玖部

《呦呦鹿鸣——燕国公主眼里的霸国》

首都博物馆学术委员会

（首都博物馆书库 编辑委员会）

主　　　任：郝东晨　郭小凌

常务副主任：黄雪寅

委　　　员：（以姓氏笔画为序）

龙霄飞　冯　好　刘树林　孙五一

吴　明　武俊玲　武望婷　钟　梅

徐　伟　章文永　鲁晓帆

《呦呦鹿鸣——燕国公主眼里的霸国》

山西省考古研究所

山西博物院　　　　　编

首都博物馆

考古发掘：谢尧亭　王金平　杨及耘　李永敏

摄　　影：厉晋春　秦剑

英文翻译：张贵余　赵雅卓　邵欣欣　杨丽明

图书在版编目（CIP）数据

呦呦鹿鸣：燕国公主眼里的霸国 / 山西省考古研究所，山西博物院，首都博物馆编. —北京：科学出版社，2014.7

ISBN　978-7-03-041415-1

Ⅰ.①呦…　Ⅱ.①山…②山…③首…　Ⅲ.①周墓－出土文物－研究－翼城县－西周时代　Ⅳ.①K878.84

中国版本图书馆CIP数据核字（2014）第158173号

责任编辑：张亚娜　宋小军

特约编辑：杨　洋

责任校对：彭　清

责任印制：赵德静

装帧设计：北京美光设计制版有限公司

出　　版：科学出版社

地　　址：北京东黄城根北街16号

邮　　编：100717

发　　行：科学出版社发行　各地新华书店经销

印　　制：北京华联印刷有限公司

印　　次：2014年7月第 1 版　第 1 次印刷

印　　张：25

开　　本：787×1092　1/8

定　　价：228.00元

（如有印装质量问题，我社负责调换）

展览项目组

呦呦鹿鸣

燕国公主眼里的霸国

Harmonious Life: The State
of Ba in the Eyes of a Yan Princess

展 览 策 划：杨文英　石金鸣

项 目 主 持：杨文英

项 目 统 筹：龙霄飞

展 览 协 调：鲁晓帆　梁育军

学 术 指 导：谢尧亭

艺 术 总 监：徐　伟

展 览 责 任 人：孙　珂

展 览 大 纲：谭晓玲

资 料 编 辑：郭喜锋

展 览 设 计：李丹丹

布　　　展：张贵余　索经令　黄雪梅　孙秀清

　　　　　　徐　涛　夏天龙　王　超

呦呦鹿鸣 霸国

燕国公主眼里的霸国

Harmonious Life: The State
of Ba in the Eyes of a Yan Princess

目 录

Contents

致 辞

山西省考古研究所所长
谢尧亭

呦呦鹿鸣
燕国公主眼里的霸国
Harmonious Life: The State
of Ba in the Eyes of a Yan Princess

　　首都博物馆以优秀的团队、良好的社会服务和得天独厚的区位优势位居业界翘楚，山西省临汾市翼城县大河口霸国墓地的重要出土文物能够在首都博物馆与公众见面，是一件非常荣幸的事情，我谨代表山西省考古研究所对这次展览的成功举办表示衷心的祝贺！

　　西周时期周王朝分封了很多诸侯国，燕国、晋国、齐国、鲁国等大国见于文献记载，还有很多小国不见于记载，像近年在晋南地区发现的绛县横水的倗国和本次展览的翼城县大河口的霸国就属于后者，特别重要的是在大河口墓地 1 号大墓中发现了多件燕侯旨的青铜器，其中的铭文说明燕国和霸国之间可能存在联姻关系，燕国与霸国之间的这种交往关系为本次展览的成功举办提供了历史的契机。通过霸国墓地出土的文物我们还可以看出，这个小国曾经强盛一时，它与周王室和其他诸侯国存在着千丝万缕的联系，它的发现对于研究西周时期的历史文化具有重要的意义。

　　最后，祝愿本次展览能够给首都人民和来自国内外的各界朋友带来文化的享受，祝愿本次展览取得圆满成功！

Address

Xie Yaoting

Director of Shanxi Provincial Institute of Archaeology

As one of the top museums, Capital Museum in Beijing is famous for her good team work, excellent social services and unique geological advantages. It is a great opportunity for people living in Beijing to appreciate the treasures discovered in the Dahekou tomb complex in Yicheng County of Shanxi Province. On behalf of the Shanxi Provincial Institute of Archaeology, I would like to express my sincere congratulations to the successful exhibition held in Capital Museum.

In the Western Zhou Dynasty, there were many vassals enfeoffed by Zhou court, for example, the States of Yan, Ji, Qi and Lu. Those were the vassal-states recorded in the historical documents. Nevertheless, a number of small states could never be found in the Chinese classical literatures. I can take two examples in Shanxi, one is the State of Peng discovered in Jiang County and the other is the State of Ba. I must note that some bronze vessels belonged to Zhi, the vassal of Yan (of what is now Beijing), were found in Dahekou tomb complex. Furthermore, the inscriptions on the bronze vessels implied that some marital relationships between Ba and Yan probably existed. Thanks to the ancient marriage system, there are more reasons for the partnership between Shanxi Provincial Archaeological Institution and Capital Museum. By the way, despite of its small territory, the State of Ba had been prosperously developed in the past, which provided a channel for other large vassal states and Zhou court through diverse dialogues. We could get more information about Zhou Dynasty from the treasures of Ba.

I hope this exhibition a great success, and I believe it can bring a wonderful cultural enjoyment to people from all walks of life.

致 辞

山西博物院院长
石金鸣

呦呦鹿鸣
燕国公主眼里的霸国
Harmonious Life: The State
of Ba in the Eyes of a Yan Princess

山西是煤炭资源大省，长期以来为国民输送了无数的光明与温暖；山西是文化遗产大省，百万年积淀为我们留下了诸多历史建筑和文物珍品。山西现存 452 处不可移动国宝，是国家级重点文物保护单位最多的省份；山西博物院的"晋魂"基本陈列，荟萃了近百年来田野考古的重要发现与研究成果，是公众了解山西地域历史文化的重要窗口。即将与大家见面的《呦呦鹿鸣——燕国公主眼里的霸国》特别展览，就是山西"晋魂"系列展示的最新考古发现与研究成果。

汗牛充栋的史书往往只是历史冰山的一角，更多的文化碎片隐没在漫漫黄土中。考古学家对考古调查、考古发掘的遗物、遗迹，以及它们的相互关联都有着浓厚的兴趣，以此来分析、探索遗物本身的信息，解释某一重要历史事件发生的过程。大河口西周墓地的发现表明，田野考古再一次发挥了它"正经补史"的重要作用，一个史书阙载的古国——霸国逐渐被世人所知。丰富的出土资料展示了霸国独具特色的文化，为揭开有关霸国的一系列谜团提供了重要线索。

《呦呦鹿鸣——燕国公主眼里的霸国》，以山西省考古研究所主持的大河口墓地发掘和初步研究成果为基础材料，为公众展示了西周早期霸国的重要考古成果，并以霸伯随葬提梁卣的铭文："燕侯旨作姑妹宝尊彝"寥寥九言为素材，讲述了霸国与燕国间一段浪漫动人的联姻故事。美丽的燕国公主，即召公奭的女儿、燕侯克的妹妹、燕侯旨的小姑姑，远嫁千里之外的霸伯。燕侯旨为他的小姑姑精心制作了成套的青铜器陪嫁贺礼。文献中不曾见燕国与霸国政治外交关系的记载，公主的夫君也不知何故未到五十而英年早逝。孤寂凄楚的长夜灯影相思泪，只好留给文学爱好者去续写了。

首都博物馆以其宏大的建筑、丰富的展览、先进的设施、完善的功能，早已跻身于国内一流。国际先进的博物馆行列。北京市文物局和首都博物馆确定并策划了此次特展，实现了山西古代文明展走进首都博物馆的多年夙愿。感谢以首都博物馆为主的展览课题组人员，他们为展览付出了诸多辛劳与智慧。预祝展览取得成功，相信此次展览会得到爱好历史与文化的北京观众的关注和喜欢，同时也期待着两馆之间继续开展更多的展览合作与业务交流。

Address

Shi Jinming

Director of Shanxi Museum

As is well known, Shanxi is one of the largest coal producing provinces, which bring light and heat for many Chinese people. At the same time, Shanxi is also famous for its cultural heritage, including numerous historic buildings and cultural relics. There are 452 unmovable heritages in Shanxi and the most national key cultural relic protection units among Chinese provinces. The permanent exhibition of Shanxi Museum displays the important archaeological discoveries and scholarship and is also the key window of the history and culture in Shanxi for visitors. The forthcoming special exhibition is just the one section of this permanent exhibition according to the latest archaeological discovery of the State of Ba in Shanxi.

The ancient history records, despite of their large amount, is always like the tip of the iceberg. And the excavated objects contain the supplement of the lost information in those records. During the archaeological investigation and excavation, the archaeologists deal with mass information, such as data of the cultural relics, ruins and the relationships between them. By means of the information, archaeologists can explain some events in the history. The discovery of Dahekou tomb sites of the Western Zhou Dynasty exactly suggests the important role of archaeology on recognition of the past. Because of this archaeological excavation, the State of Ba, which is never referred in historical records, is gradually known by people. The rich objects unearthed from the sites present a unique culture of the State of Ba and provide more useful clues to the unknown history of the State of Ba.

The exhibition *Harmonious Life: The State of Ba in the Eyes of a Yan Princess* reflects the research result of the archaeological excavation of Dahekou tomb sites under Shanxi Provincial Institute of Archaeology. Among the funeral objects, an over-top handled *you* pot has the inscription recording the marriage between the states of Ba and Yan. The beautiful Yan princess, that is, the daughter of Duke of Shao, the sister of Vassal Ke and aunt of Vassal Zhi, got married with the monarch of Ba. Zhi, vassal of Yan prepared a set of exquisite bronze vessels as the dowry. The ancient literatures don't mention any diplomatic or martial relationship between Yan and Ba, and we still don't know many details of the history such as the reason why Yan princess's husband died before fifty years old. Maybe we can leave the romantic story to the novel writers.

The Capital Museum with its magnificent architecture, rich exhibition, advanced facilities and perfect services, ranks among the first-class Chinese museums. Beijing Municipal Administration of Cultural Heritage and Capital Museum planned this special exhibition and help us to realize the dream of promoting Shanxi ancient civilization in the Capital Museum for many years. Many thanks to the work team of this exhibition and their hard work. I believe that people who love history and culture will enjoy the exhibition. Furthermore, I look forward to establishing more cooperation and partnership between both of our museums. At last, I wish the exhibition a complete success.

致 辞

首都博物馆馆长
郭小凌

呦呦鹿鸣
燕国公主眼里的霸国
Harmonious Life: The State
of Ba in the Eyes of a Yan Princess

　　人类历史很大程度上是一种不断纠错的历史。在历史学领域，史学家们在不断改写专题、国别、地区和世界的历史。这种改写自19世纪以来越发频繁与准确，因为新的考古发现以难以置疑的一手实证不仅填补文献记载的空白，而且一再纠正甚至颠覆着文献记载的内容。比如人类体质进化与社会进化的历史，再如从无文字记载的苏美尔文明、哈拉巴文明、赫梯文明等的发现。这样的改写在世界各地几乎每天都在进行。它告诉我们，人类对过去的自我认识永远是动态的，从来没有一劳永逸的"终极的历史"。

　　2007年5月，一个早已湮没无闻的地域小国或封国——霸国在山西省临汾市翼城县大河口面世。这是考古发现填充先秦史空缺的又一个例证。考古学家们在大规模的西周墓葬群中发现了风格独特的葬俗，数量惊人的随葬品，包括漆木器、青铜器、原始瓷器。有几件与北京史有密切关联的青铜器，其中一件上载"燕侯旨作姑妹宝尊彝"铭文，显然这是燕侯旨的小姑姑的器物。它的背后含有太多的未知，如燕国公主的用品为何到了这里？是否燕国公主嫁到霸国？倘若答案肯定，那么她嫁给了何人？她的婆家霸国为何不见记载……

　　北京地区有燕国的故都，可谓是燕国公主的娘家。一个佚名的贵族女子穿过太行山，与另一国的男子结为夫妻。这个多半是历史事实的佳话使不见经传的霸国与北京建立了必然的联系，使首都博物馆收藏的燕国遗存有了新的比照对象，也使这批山西出土文物在北京有了展示与研究的价值。这正是即将在首都博物馆开展的《呦呦鹿鸣——燕国公主眼里的霸国》想要告诉观众的信息。

　　为了办好这个展览，首都博物馆的策展人员进行了精心的构思，把"礼"作为连接霸国与燕国的纽带，试图重构燕国公主的出嫁过程，再现西周时期的婚礼、祭礼、丧礼、宴礼的情状，并以霸国墓葬出土文物展现文献失落之国——霸国的礼仪文化，以及由联姻产生的两国间的友好关系。

　　展览得到山西省考古研究所、山西博物院的大力支持。它的成功举办是三家单位精诚协作的成果。借图录出版的机会，我代表首都博物馆向山西同仁表示由衷的感谢。愿观众们欣赏与喜欢这个展览。

Address

Guo Xiaoling

Director of Capital Museum

The history of human beings is, in a way, a process that people keep correcting errors by themselves. The historians always rewrite the history of the world, the nation, the locale or some special topic due to new historical discoveries. The rewriting of history becomes more and more comprehensive which has frequently happened since 19th century. The reason is quite clear that the incredible archaeological discoveries, as the primary sources, correct and even overthrow the historical data documented by previously literatures; for example, the history of anthropology regarding to the physical and social evolution and the discoveries of Sumer, Harappa and Hittite which never appeared in the historical sources. This sort of rewriting history is happening everyday around the world. It suggests that the eternal history never exists, all history, as Benedetto Croce said, is contemporary history.

The discovery of the State of Ba, a small vassal state of Zhou Dynasty in Dahekou tomb sites in Yicheng County, Shanxi Province in May 2007, suddenly came into the public view. It is a perfect example of filling the vacancies on Chinese pre-Qin history by archaeological discoveries. At the tomb sites, archaeologists discovered many funeral objects, including lacquer, bronze and proto-porcelain wares, one of which suggests the close relationship between Ba State and Beijing in ancient times. Among the unearthed objects, one special bronze vessel inscribed with "Yan Hou Zhi Zuo Gu Mei Bao Zun Yi" was identified that it was made for the aunt of Zhi, the Marquise of Yan. There are still a lot of stories behind, which haven't been revealed to us so far. For instance, why the commodities used by the Yan princess was transported to the State of Ba. If there was certainly a marriage between the State of Ba and Yan, and then with whom the princess was wedded? Why is there no historical data about the State of Ba recorded in the ancient literatures?

Beijing had ever become the old capital of Yan State, her parents' home, from where Yan princess was born. She turned herself into the bride to the State of Ba linking the unknown state with the Beijing area. The comparisons between the treasures collected in the Capital Museum and the discoveries found from Shanxi are the best references available for display and study on the history of the Yan State. That's what we present in *Harmonious Life: the State of Ba in the Eyes of a Yan Princess*.

We try to create curatorial plan for restoring the process of the princess's wedding ceremony back to the original as well as for bridging Ba and Yan states through traditional ritual behaviors popularized during the Western Zhou Dynasty at the worshipping, funerary and banqueting spots. The exhibition demonstrates the ritual culture of Ba State from the objects unearthed in Dahekou tomb sites and the friendly diplomatic relationship existed between Yan and Ba by marriage.

The exhibition is benefited by the fruitful cooperation with the Shanxi Provincial Institute of Archaeology and Shanxi Museum. On the behalf of Capital Museum, I express my heartfelt thanks to them for their generous help and efforts. At last, we hope our visitors enjoy this exhibition.

呦呦鹿鸣

燕国公主眼里的霸国
Harmonious Life: The State
of Ba in the Eyes of a Yan Princess

解 读 霸 国

谢尧亭

引　言

　　霸国是根据 2007—2011 年发掘被盗的山西省临汾市翼城县大河口西周墓地时出土的青铜器铭文而确定的一个西周诸侯国。此前没有人知道中国历史上还有一个霸国，浩瀚的史书中没有留下关于它的只言片语，那么这个神秘的国家如何能够隐藏了近三千年而无人知晓？它又是怎样的一个国家？存在了多长时间？它来源于哪里？最后又去了哪里？……太多的问题需要我们慢慢地去揭开她神秘的面纱。

　　中国有五千年的文明史，也就是说从今天的 2014 年，往前追溯四千年，才到我们传世文献记载的夏朝，即公元前 21 世纪。那么我们从夏朝再往前追溯一千年，即公元前 31 世纪才能够得上五千年的文明。一般来说，严谨的学者将夏代及其以前都归为传说时代，而将商代后期发现甲骨文字以后的历史作为信史。为了给研究中国五千年文明史创造条件，20 世纪 90 年代我们国家启动了夏商周断代工程，后又启动了中华文明探源工程，旨在通过多学科合作的手段主动探索、研究和排定中国夏商周时期的确切年代。而霸国的横空出世，极大地丰富了中国五千年文明史。

历史背景

　　公元前 1000 年前,商王朝统治着中原大地,和它同时期并存的还有大量的国家。这些国家，有的是商王朝分派出去的贵族建立的据点，有的是与商王朝关系密切、结成联盟的方国，还有的方国时叛时服，更有一些方国与商王朝长期对抗。这些国族大都是从古代的部族发展过来的，由于地理环境的阻隔，交通和通讯的落后，这些国族的生活方式与中原以农业为主的国族在经济形态、政治结构、文化和军事等方面形成了一定的差异，中原国家往往将这些与其自身文化不同、经济形态有异的族群称为夷狄或戎狄。总之，在当时的中国大地上分布着很多的族群或国家，具体数量已不可而知，单从后世文献记载来看大体上应有数千个这样的国族。

商代末年，商王纣征伐东方的部族过于频繁，导致老百姓怨声载道，兵士苦不堪言，国力大减。而西北方有戎狄等国族的不断侵扰，西土的周族实力渐强，周武王联合所谓的八百诸侯在商郊牧野一战消灭了商朝。周族或周邦，本是商王朝统治时期位于西方的一个小国家，文献上称为"小邦周"，其文化和经济较商王朝落后，但周族的首领励精图治，奋发图强，以新生力量摧枯拉朽、推陈出新，夺得中原领袖的大权，面对商王朝及其属国、方国等大片江山，周族统治者采取了封建诸侯的方式以"藩屏周室"，即保卫周王朝的江山。周王分封了儿子、兄弟或功勋卓著的大臣，建立了齐、鲁、燕、晋、郑、卫、荀、芮等诸侯国，还分封了先圣王尧舜及夏商的后代，同时还将商王朝原来的旧国重新分封，甚至将很多戎狄族群也进行了迁封，这样在周王朝统辖的区域内有 1700 多个分封国。这些分封国，说是国家，其实都不是很大，大者有方百里，中者有方五十里，小者有方三十里，也有更小的。当时的自然环境与今天不同，人口稀少，气候温暖，丛林茂密，道路交通也不像今天这样发达，因此，在一个区域中一般都会安排有一个大国作为区域的方伯国，这个方伯国一般不干涉其他小国的内政，各个国家一般都有相对独立的处理内政外交的主权，但是在给周王室缴纳贡赋和对抗非华夏戎狄族群战争方面要听从方伯国的组织和召唤，这个方伯相当于二级行政区官员，在某种程度上替周天子治理一方，但事实上他又没有多少实际的控制权，周天子可以越过他直接与小国对话或往来，这大概就是当时的政治体制。

西周时期特别是西周早中期周王室对诸侯国进行着有效的管控，管控的手段有分封、册命、朝聘、巡狩、征伐等，这些都通过"周礼"来规范，周礼相当于西周的法规。我们知道，在周王室的统治范围内，青铜器和玉器等具有相同或相似的器形、纹饰、制法，这些器物的资源若非周王室直接控制，就会存在多种多样的风格，当然这种管控并不意味着所有青铜资源和玉器资源的原始加工和深度加工都要在周王室完成，比如铜矿的拣选、青铜的冶炼等都在矿料产地附近进行，甚至周王室也可以在青铜产地附近设置作坊进行铸造。目前所知西周时期的铸铜作坊在丰镐、周原和洛阳都有发现，其他封国也有少量陶范零星发现，似不足以说明其拥有铸铜工业，比如在横水墓地的墓葬中就发现过陶范，它作为随葬品出现在被盗掘的大墓之中，可能是赠送或外来之物，在天马—曲村遗址 J7 区发现的陶范，其时代主要是春秋早期，有少量陶范《天马—曲村》作者认为可能为西周晚期之物，笔者在《天马—曲村》报告中没有找到发表的与陶范共存的陶器，而且这批陶范与西周晚期的青铜制品的风格也不符，因此这些陶范的年代可能也要晚到春秋早期。玉器资源也一样，从矿料的粗加工到玉器的深加工，从矿源到矿工和玉器工艺师都被周王室垄断，各个诸侯国家之所以能够拥有这些制品，一种方式是通过分封赐器，另一种方式是册命赐器，还有军功赐器、往来朝聘巡狩赐器，甚至通过赗赙、赠送、媵器等多种礼仪交流的方式来实现器物的流通，当然，还有一种主要的获取方式，就是诸侯国根据需要来订做，通过用海贝、马匹、丝帛等等价物品交换来获得，像原始瓷器、金器、海贝等物品都是外来品，这些东西当然不是订做的。西周的青铜器和玉器等资源都被周王室掌控着，一个国家拥有这些资源和财富的多少表明其国力强盛的程度，一个贵族拥有这些资源的多少则表明这个贵族地位和财富的拥有能力的程度，因此，这些用来表达财富、身份、地位的重要资源在礼制比较严格的西周社会周王室必须进行牢牢地把控，西周早中期周王室也完全具备这种条件和能力，到了西周晚期特别是春秋早期，随着王室的衰落，各诸侯国不再把天子太当一回事了，因此各地都纷纷筑炉起灶，铸铜治玉，各诸侯国的文化特色立即凸显了出来，这是西周与东周政治体制不同带来的直接后果。春秋时期各诸侯国又复习了一遍西周王室的课文，走了一遍西周王朝的老路，各诸侯国进行国内分封，到战国初年，"诸侯倒霉了，卿大夫起来了"。历史在西周和春秋两个时期以相似的版本不同的人物重新上演了一遍，只是大舞台换成了小舞台而已，让诸侯们重温了一遍周天子的悲剧，可见西周以来的分封制存在先天的弊端。著名的历史学家吕思勉先生认为封建有四次反动，一次是项羽的复辟，一次是刘邦的封建，一次是西晋的分封，一次是朱元璋的封建，结果均以失败而告终，所以吕先生说"封建之反动，实至第四次而终"。

确认霸国

霸国是怎么确定的呢？换句话说，怎么知道这里有个国家叫霸国呢？"霸国"一词是 2007 年由我们大河口考古队根据出土青铜器铭文最先叫出来的。2007 年 5 月在位于翼城县县城以东约 6 公里的大河口村北高地上发现盗墓，经报请国家文物局批准，山西省考古研究所 2007 年 9 月对这里正式进行抢救性考古发掘。大河口墓地面积 4 万多平方米，埋葬有一千多座墓葬。在大河口 2 号墓中的一个铜甗的内壁残片上最先发现了青铜器铭文，内容是"唯正月初吉？伯作宝甗"，关键的是铭文"伯"前面的这一个字，我们都不认识。这个字左边是个木字旁，右边写得有点奇怪，查找相关字书也没能辨认出来。有人说我们考古工作者天天盼着出土的青铜器上有铭文，可真正发现了铭文，关键的字却又释读不出来。索性就先存疑，等更多的相关考古材料发现吧。到了 1 号墓随葬的青铜器暴露出来的时候，可想而知我们当时急迫的心情。在一件大鼎的内壁，我们发现了铸刻得很浅的"伯作宝鼎"四个字，只是这还无法解决关键性问题。不过我们由此知道这座墓葬的墓主人是伯一级的贵族，即国君或族长。使我们惊喜的是，在一个铜簋的盖子内面我们发现了铸造有"霸仲作旅彝"字样的铭文，我们知道，伯、仲、叔、季是排行老大、老二、老三、老四的意思，而这个"霸"字应该就是国名或族名了。毕竟此时我们的证据还很少。之后，我们又在一本字书中查到了一个"霸"字的下部的写法与 2 号墓铜甗上的那个释读不出的字十分相似，我们由此推测 2 号墓铭文的这个？伯应该是霸伯。

有意思的是，2 号墓是一座女性的墓葬，却见有霸伯的铜甗，而没有见到与这位女性墓主有关的铭文。我们之所以判定 2 号墓墓主人是女性，主要是依据该墓随葬的物品，比如带梯形牌的串饰一般是女性的专有用品，另外就是没有随葬青铜兵器。像 2 号墓这样规模较大的墓葬如果是男性墓葬，在西周时期，不随葬兵器一般是不可能的事情。同样有意思的是，我们在 1 号墓葬中并没有见到一例带有"霸伯"铭文的青铜器，但这座墓葬却随葬了大量的青铜兵器，没有发现那种梯形牌的串饰，因此我们推断 1 号墓的墓主人是男性。同时我们在 1 号墓随葬的青铜器上还发现了"燕侯旨"的铭文，我们已知燕国在北京市房山区的琉璃河遗址，而大河口墓葬所处的这一地带绝不可能是北燕国的所在。因此我们就大胆地提出了"霸国"的发现。后来 2009 年大规模发掘的时候，在 1017 号墓的青铜器上发现了一个清晰的"霸"字，也进一步地印证了我们的推测。

霸氏归属

仅凭这个"霸"字就能称国吗？它会不会是晋国的一部分呢？其实，在此之前我们发掘绛县横水佣国墓地的时候，已经对类似的问题进行了长期的思考，绝不是贸然地、想当然地提出一个标新立异、哗众取宠的热词。学术研究是一件非常严肃的事情，它需要有长期的知识积累，特别是考古学，不仅需要一定知识的储备，还需要有丰富的实践经验和必备的理论修养。我们知道，绛县横水墓地发现了几十座大墓，其中还有三座带墓道的大墓，墓道在古代社会不是人们随便使用的东西，尤其是在礼制相对比较严格的西周，人们更不会随便使用。这些墓基本上都有独立陪葬的车马坑，大墓中大多随葬有丰富的青铜器、玉器等。从出土青铜器上铸造的铭文可以看出，佣国具有相对独立的外交和内政，它与周王室及其他国家有相互的往来关系，而且墓地的埋葬习俗显示他们的人群构成相对单纯，具有相对雄厚的实力。我们通过大河口墓地与曲沃县北赵村的晋侯墓地、绛县横水村的佣国墓地的对比，发现在这里有埋藏丰富且有超过晋侯墓随葬品的墓葬，这里的大墓也有独立的车马坑，在大墓中还随葬有原始瓷器、金器等珍贵的外来品，这些物品不是一般贵族可以拥有和随葬的奢侈品。综合以上多个方面的因素，我们才敢确定这里应是一个独立的国家——霸国。

当然，也有相当一部分学者和专家认为"佣"和"霸"不是国家，他们认为佣和霸最有可能就是文献中记载的"怀姓九宗"中的两宗，也就是分赐给晋国的始祖唐叔虞的媿姓狄人。他们认为霸氏就是晋国的一个组成部分，甚至也有专家提出佣和霸都是晋国的采邑。他们推论的基础之一是大河口墓地的霸氏在晋国国都的附近，二是这两个人群都是或可能都是媿姓，正好可以与文献所记的"怀姓九宗"联系起来，因为媿与怀本来是相通的字。乍一看，这个说法似乎有一定道理，作为晋国的一部分也颇为合理。其实不然，晋国分封时地方并不大，《史记·晋世家》记载为"方百里"，方百里的意思就是长百里、宽百里，古代的里比现在的里要小，当然分封的方百里绝对不会那么方正，但也不会漫无边际地大出很多。因为自原始社会以来，特别是新石器时代人群数量众多，从我国目前已知的各地的新石器时代遗址数量即可窥见一斑。大禹会诸侯时万国林立，到商代还有几千个国家，周武王伐商时联合了八百诸侯会盟津，西周国家多至近两千个，可想而知，当时一个国家的面积和范围有多大。如果把佣、霸作为晋国的一部分，则已超出"方百里"的范围了，更何况有几宗怀姓目前还没有发现。再者是在晋都翼，即天马一曲村遗址就发现有头向西的媿姓狄人的墓葬，其中也有贵族，这与《左传》上记载的"怀姓九宗"在翼都是相一致的。因此，笔者认为佣和霸应是拥有独立主权的小国，即便是像有些人提出的它们是附庸国，也必须承认它们是国。更何况西周时期的附庸国与后世的附庸国含义有很大的区别。

霸国与晋国为邻，它们之间有着密切的关系。2002年在香港市场上发现了一件晋侯铜人，其胸腹部有铭文"惟五月，淮夷伐格，晋侯搏戎，获厥君冢师，侯扬王于兹"，晋侯铜人铭文的发现说明了晋救霸难。在曲村墓地6197号墓葬发现过一件霸伯簋，这座墓葬的墓主为女性，霸伯簋发现于这座女性墓葬中绝非偶然，因此有人推测这位女性来自霸国，是霸伯之女，她的丈夫是6195号墓主，这两座墓葬都是东西向墓葬，墓主头向东。我曾经研究认为曲村墓地头向东的贵族是土著唐人的后裔，当然唐人也不是一个单纯的人群，应是以祁姓唐人为主体的可能含有多个姓氏的族群集团，像祁姓范氏和嬴姓赵氏在晋国早期可能都属于唐人集团。这件霸伯簋可能是作为陪嫁的嫁妆埋葬到霸伯女儿的墓葬中。这说明晋、霸两国族通婚，霸女嫁给晋国的唐人贵族而不是嫁给晋国国君，当然我们不排除也有霸国公主嫁给晋侯的可能，从燕国公主嫁到霸国来看，这种可能性是完全存在的，只是目前还没有发现明确的证据而已。另外，传世的青铜器——佣生簋的铭文记述了佣生与格伯（即霸伯）以马匹换田地的故事，从铭末的族氏铭文看，佣生当是妘姓瑅人，佣生即佣的外甥，佣生的母亲是佣国人，佣生所在国与霸国为邻，若不排除妘姓瑅人是晋国人的一

14

部分，那么佣生完全可能是晋国人，除非在霸国周围还存在一个琱国。晋国和霸国的密切关系还体现在考古学文化间的一些相似点，例如车马坑都位于主墓以东，车马坑为东西方向，没有殉人和俯身葬，人骨的种群特征更为接近等。但遗憾的是，到现在为止还没有在大河口墓地发现明确的与晋国有关的文物。

大河口墓地的青铜器铭文显示这里有一个霸氏族群，首领是霸伯。我们在大河口墓地看到更多的是周文化的影响和独立发展的印迹，它有独立的物质文化和精神文化，拥有独立的政治、经济、外交和军事，它有自身的信仰意识，这样的族群就是西周的一个封国——霸国。笔者认为西周时期的采邑政治在各诸侯国是不存在的，那是东周时期发生在诸侯国的事实，因此，把霸国和佣国说成是晋国的采邑根本就是讲不通的，把这些小国说成是晋国的附庸国也没有什么道理。其实晋国作为晋南地区较大的侯国，可能是方伯国。

燕国公主

大河口1号墓中出土了多件与燕国有关的青铜器，其中有一件青铜卣的盖内和器内底铸造有相同的铭文"燕侯旨作姑妹宝尊彝"。燕侯旨是燕国的第二任国君，这件器物应是他为他的小姑姑做的青铜宝卣。我们知道，以往发现的燕侯旨青铜器十分罕见，仅在日本和上海的博物馆有存，因此燕侯旨的青铜器十分珍贵，这件卣内还发现了成套的酒器七件，它们分别是大小各异的五件饮酒的觯，还有一件舀酒的斗和一件单把的小罐。这种形制的小罐，以往没有发现过，器身上有几何形的奇特的纹饰，从与其他器物一起发现于卣内的关系推测，这个小罐可能也是舀酒用的。但令人不解的是，它的把较宽，而且与器身间距较小，手指难以穿过，但它又不像是明器，难道就是手指捏在把的两侧边来使用？或者另有木制的柄嵌于鋬手内当做斗来使用？或者是用来饮酒用的？它的学名又叫什么呢？这些问题现在还回答不了。另一件器形和纹饰相同但略小的铜卣，由于盖子与器身锈在一起，还没有打开，但通过X射线拍摄可以看出，在这件卣内也有一套酒器。此外，在两件爵、一件尊和一件觚上也发现有与燕侯旨相关的铭文，这些发现大大地丰富了燕侯旨的青铜器，并且把燕国和霸国牵连到了一起，这在大河口墓地发现之前我们是不敢想象的。

关于燕侯旨是燕侯克的儿子还是弟弟，在史学界还没有定论。《史记·燕世家》说封召公奭于北燕，唐代司马贞的《索隐》认为召公奭并未就封，而是让其大儿子去代父就职，以其二儿子留在周王朝继召公奭之位。但文献中都没有记载其长子、次子的名字，而且《史记》对召公奭以下九世均没有记载，因此，燕侯克和旨的青铜器铭文至为珍贵。克和旨是两代燕侯没有问题，只是二人若是兄弟，则燕侯旨的小姑姑则是召公奭的妹妹；二人若是父子，旨的小姑姑则是召公奭的女儿。从大河口墓地 1 号墓的年代来看，似乎"姑妹"为召公奭的女儿、旨是克的儿子较为合理。从相关的青铜器铭文来看，"召伯父辛"应该指的就是燕侯克。1 号墓的年代为西周早中期之际，即昭王和穆王之际，而大部分青铜器包括燕侯旨的器物年代都在西周早期，只有个别青铜器形制指示的年代略晚，也就是说 1 号墓葬的墓主霸伯埋葬的时间在西周早中期之际。西周早期有 90 年左右，假如霸伯活了60 岁，他 20 岁结婚，那么他结婚时应在西周早期成王和康王之际，这与燕侯旨等器物的年代大致是吻合的，作为召公奭的小女儿结婚的年龄也是可以讲得通的，但若是召公奭的妹妹则显得有些滞碍不通了。

那么我们怎么知道是燕国的公主嫁到了霸国呢？首先我们确信霸国不是姬姓国家，主要原因是大河口墓地墓主头向以向西为主，还有很多腰坑与殉狗，当然有人认为霸就是格，霸为姞姓，有人认为是媿姓，以后者为主流意见。燕为姬姓，异姓通婚，是当时的常制，最重要的证据就是在 1 号男性墓葬中出土了多件套燕国国君"旨"的青铜器，铭文中显示为燕侯旨为他的小姑姑制作的卣、尊及燕侯旨作的爵和觚等。大量的燕侯作器埋藏在霸国国君墓葬中，这不是一般的助丧之器（赗赙）所能解释的，更不可能是分赐、掠夺得来的，赠送的唯一途径可能就是两国联姻，燕侯旨给他的"姑妹"（小姑姑）专门制作的青铜礼物，赠送而来。当然，这个赠送行为可能发生在结婚的时候，也可能发生在婚后某一时间，比如有什么值得庆贺的喜事。反过来，如果不是燕国公主嫁到霸国，这种特殊铭文的成套青铜酒器应该很难流传到这个异族小国。所以，我们推断这两个国族联姻是可以令人信服的。但遗憾的是到现在为止我们还没有发现或确认燕国公主的墓葬，一是因为大河口墓地还没有全部发掘，二是大河口墓葬极少见夫妻并穴合葬的现象，三是我们发现的女性大墓墓主族姓身份均没有铭文可以确定。例如大河口2号墓葬就是一座大墓，墓主为女性，但青铜甗上的铭文是"唯正月初吉格（霸）伯作宝甗"，未见与女性墓主有关的文字，同样在1号墓葬中也没有见到与霸伯夫人有直接关系的文字，比如族姓或名字，或她制作的器物。奇怪的是燕侯旨为其姑姑作的器物为什么不埋藏在他姑姑的墓葬中，而埋到了霸伯的墓葬中，而且燕侯旨为什么要给他的小姑姑赠送一批酒器而未见其他器物呢？难道说他的小姑姑喜欢饮酒不成？事实可能并非如此，在西周早期赠送酒器或者是一种礼仪，是否与谐音"久"字有关，表示长久之意，我们并不能证证，不管怎么说，燕国的这批青铜酒器一定有其特殊的含义。

如果说酒器在西周早期还占有一定重要地位的话，它们埋葬到国君霸伯的墓葬中而没有埋葬到其夫人燕国公主的墓葬中，可能更为合理，也更能使人理解。从这个意义上来说，霸伯国君的地位显然要高于其夫人，而不像绛县横水西周墓地1号墓与2号墓那样，夫人的1号墓规模大、随葬器物丰富、级别较高，而倗伯的2号墓规模小、随葬器物少、级别不及其夫人。有人认为这可能与这位夫人的出身有一定关系，她来自毕国，因为毕公是西周王朝的重臣，这位毕国公主嫁到小国倗国，享有特别的荣宠。不过这种推测并不能使人信服，因为绛县横水西周墓地的另一位倗伯夫人还是周王的姐姐（王姊），而她的墓葬在各方面似乎并没有超过倗伯。因此，笔者认为这位毕国公主嫁给倗伯以后，在一定时期曾经主持过倗国大政，拥有国家权柄，她极有可能当过倗国的国君，极有可能是目前所知中国文明史上第一位女君主，这是很了不起的一件大事，尽管倗国只是一个小国，甚至有人认为可能是晋国的附庸。

在大河口2002号墓葬中发现了一套水器盘盉，这套水器上各铸有一长篇铭文，其中这件青铜盉的样式为鸟形盉，这种形制属于首次发现，在其背部有一个青铜链盖，盖内有51个字的铭文，内容讲的是"乞"这个人向主人发誓如何如何。从铭文自名为盘盉来看，我们确定这件青铜器为盉，同出的还有一件青铜盘，盘内也铸有一长篇铭文，内容也是说乞誓的事，其中多次提到霸姬，这位霸姬肯定是2002号墓主霸伯的夫人，她来自晋国还是燕国？抑或芮国等其他姬姓国家？我们不知道，但我们知道2002号墓主是一位男性，是一代霸伯，从霸伯与姬姓国家联姻来看，霸国的族姓一定不是姬姓。

漆木人俑

在 1 号墓葬中发现了大量的青铜器、玉器、漆木器等贵重的文物，特别值得一提的是两件漆木人俑。这两个人俑可以说是目前中原地区发现的年代最早的人俑，但事实上他们并非殉人的替代品，他们站立于墓主人的脚端方向椁室外面的二层台上，面朝向墓主人的方向，双手的姿势像是原来手里握持着什么东西，发掘时并未发现，重要的是两个人像的脚下各踩踏着一个漆木的乌龟，造型逼真，这种现象在以往似乎没有发现过。我们知道商周时期占卜经常用龟的腹甲和背甲，在有的大墓中还随葬有完整的龟甲，文献上记载为"宝龟"，由于这种动物的寿命比较长，古代的人一般认为它阅历丰富，具有某种先知先觉的神性，或具有通神的功能。既然两个漆木人像站立于乌龟之上，一定具有某种特殊的象征意义，应与当时人的宗教信仰有一定的关系。这两个人像或者就是通天地的巫觋，在两个人像的旁边随葬有其他的漆木器，可能都与这两个巫觋有关。把他们埋葬于霸国国君的墓葬中，就不能仅仅理解为陪葬的人俑，也许与接引墓主的灵魂升天有关，或者具有佑护墓主攘除其他鬼神干扰的作用，或者具有其他的意义，还需要进一步的研究。

在大河口墓地没有发现一个殉人，这与绛县横水墓地发现大量殉人的现象截然不同，殉人是一种原始野蛮的宗教信仰的表现，自新石器时代以来特别是商代广泛流行，但周人普遍不使用殉人，这反映了宗教信仰和意识形态方面的文化差异。从这个意义上来说，大河口墓地霸国人群受周文化影响或更强烈，文化习俗上部分有别于绛县横水西周倗国墓地。殉人这种陋习沿用时期较长，流行范围较广，像三晋的赵国就流行这种习俗，太原发现的春秋战国之际的赵卿墓就殉葬了四个人，长子县牛家坡 7 号战国墓也殉葬了三个人，春秋晚期秦国的秦景公大墓中殉葬了一百数十人，侯马乔村战国中晚期围沟墓中也发现有大量殉人，到了秦汉时期，殉人恶习才逐渐销声匿迹，更多的是用陶制的人俑代替了活人，这应该说是人类历史上的伟大进步。在当时当地，给墓主殉人一定有其合理的一套说辞，具有不可抗拒的必然的理由，在一定程度上殉人这种现象与人群或主体人群推行的文化价值观有关，长期存在并根深蒂固的原因是某种文化价值观的束缚与禁锢，大河口墓地不殉葬人这种现象就说明它与横水倗国墓地在宗教信仰观念上具有很大差异。

另外，在 1 号墓葬中还发现了大量的漆木器、原始瓷器，这些物品大多放置在墓壁掏挖的土龛中，共发现了 11 个壁龛，一座墓葬中有这么多的壁龛在西周考古上是第一次发现。近年在陕西宝鸡石鼓山西周墓葬中也发现了多个壁龛，但其中放置的多为青铜器，与大河口西周墓葬不同，大河口 1 号墓的青铜器比较集中地见于墓主头前面的棺和椁之间，根据塌落的情况推测当时器物应是放在木架子上的。其实有多个壁龛的墓葬在山西省襄汾县陶寺遗址的新石器时代晚期大墓中早就发现过。

霸国葬俗

大河口墓地的墓葬方向绝大多数是东西向，墓主头向以西向为主，少量头向东，仅有 4 座头向北的南北向墓葬。大墓墓主均是头向西，可见头向西的墓主人群是大河口墓地的主体人群，那么他们的头向西与头向北有什么特别的不同或者有什么特殊的含义？为什么大河口墓葬主要是头向西的人埋葬在这里呢？或者说他们为什么要头向西方埋葬呢？埋葬人的时候头向朝哪一个方向，最初肯定是有一定讲究的，不是随随便便挖个坑埋葬就了事的。自从新石器时代人类认识到主动处理尸体的埋葬方式以来，墓葬方向和墓主人的头向就有一定的含义，而且不同的人群可能有不同的含义。当时人们究竟是怎么想的，我们已很难知晓。但我们通过考古发掘可以看到不同的人群具有相同、相似、不同的埋葬头向，即便是相同的头向在不同的人群中也未必具有相同的含义。我们根据东周或汉代的文献和后代注疏知道，头向北的原因是古人认为北方是灵魂所归宿的幽冥之地，那么头向西与向东又怎么解释呢？大家有多种猜测，有人认为某方向是其族群起源之地或祖先所在地，有人认为是灵魂的归宿地，当然也有人认为与山水地理等环境有关，但在横水、大河口和北赵几个大墓地都具有北方为山、南方为水的地貌特征，墓主头向却并不全同，即使是在同一墓地也存在各种不同的方向，这恐怕是不能用地貌特征的理由所能解释的。我们知道风水的观念是后来才有的，西周时期并没有成熟的风水理论，当然周代埋葬也需要"筮宅"，即在这规划的公共墓地中，在其家族的范围内找到其最佳的墓位。虽然我们已经研究出大河口墓地头向的不同具有划分人群的意义，但头向表示的具体含义还有必要进行深入的研究，比如搞清这些人群的迁徙和源流关系，对于探讨其头向的象征意义具有十分重要的作用。

有意思的是，在大河口墓地的国君夫妇并没有埋葬在一起，既不同穴合葬，也不异穴并列合葬，或者叫做并穴合葬，俗称"对子墓"，这与同时期的曲沃县北赵晋侯墓地的晋侯夫妇墓葬、曲沃县曲村墓地贵族夫妇的墓葬，以及绛县横水倗国墓地的倗伯夫妇墓葬等大不相同，在大河口墓地较难以确定国君霸伯与其夫人的对应关系。

大河口墓地的所有墓葬都没有发现墓道，都是竖穴土坑墓。我们知道在安阳殷墟发现的商代王陵有带四条墓道的，也有带两条墓道的，西周的王陵虽然还没有发现或确定，但周原周公庙的墓葬就带有四条墓道，显然其墓主的身份和地位是很高的，可惜这个大墓被盗惨重，没有发现明确的可判明墓主人的信息，不过大多数学者推测墓主人可能是周公。当然在燕国的琉璃河墓地也发现了一座带四条墓道的墓葬，不过这四条墓道开在墓口的四角，并且都比较狭窄，显然其象征意义大于实用意义，或与"召公建燕"的规制有关？带两条或一条墓道的墓葬在西周各诸侯国墓地比较多见，像山西省曲沃县北赵晋侯墓地的绝大多数墓葬就带有一条或两条墓道，在山西省绛县横水倗国墓地共发现了三座带一条墓道的墓葬。但在时代略晚的河南省三门峡市的虢国墓地没有发现一座带墓道的墓葬，说明墓道这种埋葬形式并不普遍，不见得各国高级贵族都一定使用。但凡是带墓道的墓葬一定是级别较高的贵族，问题是有些高级贵族为什么不使用墓道这种埋葬形式呢？虢国君主没有，霸国君主没有，倗国大部分君主没有，卫国君主也没有，可见墓道也不是高级贵族墓葬的必备之物，没有墓道的大墓其身份地位未必就低于带墓道者。使用不使用墓道与下葬方式有关？还是与信仰意识有关？或者与别的什么东西有关？在历史文献记载中墓道又称为羡道，被解释为下葬棺椁葬具使用的通道，我们已知的西周墓道有斜坡式和台阶式两种，而且我们知道西周偏早期的墓道坡度较缓，偏晚期的墓道坡度较陡，也就是说，偏早期的墓道距墓室底部较远，偏晚期的墓道距墓室底部较近，很明显其实用的理由是存在的。在大河口1号墓墓口四角发现四个斜洞，没有墓道，在横水倗国墓地也有这种现象，这些带有斜洞的墓葬年代也都属于西周偏早期，它们的墓葬体量较大，墓主身份地位较高。在湖北随州叶家山虽然发现一座墓葬既有斜洞也有墓道，但墓道不具有实用性，现在大家普遍认为这些斜洞与墓主的下葬存在必然的联系。只是西周时期在各地发现的高级贵族墓葬不都设有斜洞或墓道，也就是说大部分高级贵族墓葬既没有斜洞，也没有墓道，这些墓葬的下葬为什么没有使用这些特殊的设施呢？像三门峡虢国墓地和浚县辛村卫国墓地的高级贵族墓葬就是这样，这可能与各地的埋葬方式不同有关。与西周时期各国的信仰习俗没有多大关系，墓道或斜洞的有无与礼制也没有必然的绝对的关系，但一般的中小型墓葬是绝对没有这些设施的。

在大河口墓地目前只发现1号墓葬有斜洞，其他所有大中小型墓葬都没有发现这种设施，这座墓葬是目前大河口墓地所发现的大型墓葬中年代最早的一座。当然大河口墓地还没有全部发掘，我们不排除比1号墓更早的或同时期的大墓中也存在斜洞。斜洞的设置与下葬方式有关，但斜洞究竟是如何使用的我们并没有具体可靠的证据来说明，而且为什么斜洞到后期又废弃不用了呢？像大河口1017号大墓和2002号大墓既没有斜洞也没有墓道，之后的其他大墓也没有，这可能与大河口墓地埋葬方式的改变有很大关系。从目前所见到的材料来看，这种埋葬方式的变革时间是在西周早中期之际，而曲沃县的北赵晋侯墓地从早期到晚期大都使用墓道，没有发现斜洞，不存在这种变革，虢国和卫国墓地既无斜洞又无墓道，当然也不存在这种变革，可见这种埋葬方式的使用和变革也不存在普遍性。

大河口墓地墓主人的埋葬姿势以仰身直肢为主要葬式，没有发现俯身葬，在绛县横水倗国墓地有大量的俯身葬，甚至连2号墓主倗伯都是俯身葬。我们知道，俯身葬在商文化中比较流行，等级高下不同的人群都有这种现象存在，具有某种特殊的象征意义，这说明大河口墓地人群没有这种特殊含义的信仰意识，这与天马—曲村的北赵晋侯墓地和曲村墓地是一致的，也在一定程度上说明大河口人群的特殊性，他们与横水墓地人群的埋葬习俗差异其实还是相当大的。

大河口墓地的墓葬从墓形大小、随葬品种类和数量的多少，可以看出这个霸国是一个等级明确、礼制严格的等级社会。在这个社会中，霸国国君即霸伯是第一等级，其次是霸伯夫人和其他贵族，第三等级是有随葬品的平民，第四等级是无随葬品的贫民。在大河口墓地，头向西的人群是霸国的主体人群，其他头向的人群数量较少，而且不是统治阶层，从墓葬所反映出来的贫富差距和人群结构来看，这个社会中间阶层的人数众多，富人阶层和穷人阶层的人数都较少，是一种比较合理且相对稳定的社会经济结构。因此在西周至春秋早期三百余年的历史发展过程中，霸国应该是处于一种稳定的发展状态。

大河口 1017 号大墓的棺盖板上发现了大量的海贝，这些海贝约有 2 万枚，编连有序，似乎是有一定形状的东西，笔者曾经推测它们与帷荒有关。海贝是一个总称，有很多个品种，我们所称的海贝其实是其一种。关于海贝的来源与功能很值得研究，首先我们必须承认它不是晋南本地的产物，是来自海洋的外来品，是稀有的珍贵的东西。因此，在商周时期以它作为等价交换的媒介，其实具有早期货币的功能，到西周时期它仍然充当这种角色，如金文中常见赐贝多少朋来做宝器，这个贝就是我们所说的海贝。当然它也用作帷荒上的饰品，甚至是马具络头和带饰，说明它具有装饰品的作用，它还见于墓主人的口中，作为饭含来使用。春秋时期海贝还在较多的使用，后来随着交通的发达和金属货币的出现，海贝渐渐退出了历史舞台。

大河口墓地出土的一件青铜盆（铭文称为簋）上有一篇铭文，反映了周王命令应伯征伐淮南夷的事实。西周时期淮夷与中原华夏的战争主要有两次，一次发生在西周中期，一次发生在西周晚期。自武庚叛乱以来，南淮夷与周王朝长期存在矛盾。2002 年在香港文物市场上发现的上有铭文"淮夷伐格，晋侯搏戎"的晋侯铜人，其年代大约相当于穆王时期，文献记载周穆王时徐偃王发动叛乱，周穆王的驾驶员造父驱车一日千里自西方赶回救乱，穆王后将造父封于赵城。这个徐偃王就是淮夷一部徐国的首领。晋侯铜人记载，淮夷这次进犯中原一直打到了大河口一带，晋侯奉王命与淮夷战斗，以救霸国。过去我们不知道格和霸这两个字是可以通用的，大河口青铜器上大量的铭文显示出，格与霸是一回事，格国就是霸国。西周时期周王朝四邻蠢蠢欲动，特别是西北方的戎狄和东南方的淮夷，以及江汉一带的部分族群长期与周王朝处于敌对状态。西北方有山戎、猃狁、犬戎等各种戎狄，淮夷或淮戎是对淮河流域中下游许多小国的统称，徐国在这一区域是一个大国，也称徐戎。周王朝与四方非华夏族群在文化上有区别，在生活方式上也不相同，政治结构与经济形态也不同，军事方式上也有较大区别，因此在资源、土地等方面不可避免会发生矛盾，而且这种矛盾是不可调和的，因此彼此之间的战争也是不可避免的。西周时期之所以分封这么多诸侯，特别是分封诸侯大国都具有一定的军事目的，那就是为了"藩屏周室"。从大河口出土的这件青铜盆铭文看，霸国参与了这次征伐淮南夷的战争，这次战争发生在西周晚期，结合晋侯铜人的铭文来看，淮夷与霸国之间也进行了数次战争。可见霸国早已自视为华夏族群的一部分，它早已被周王朝同化了，因此在大河口墓地除了头向、墓向、腰坑、斜洞等特殊现象以外，其墓葬的形制、葬具、葬式、陶器、青铜器、玉器等与周文化别无二致，虽然霸国与倗国还是媿姓狄人，但早已华夏化了。

霸国源流

大河口墓地所属的文化，与绛县横水墓地文化比较接近，而与以天马—曲村墓地为代表的天马—曲村文化差别较大。通过研究我们知道，大河口墓地人群可能也与横水墓地一样，属于媿姓狄人，只是二者属于狄人的不同支族，因此他们之间存在这样或那样的差异也是可以理解的，那么大河口人群来自哪里，最后又去了哪里呢？

在大河口遗址和墓地没有发现比西周更早的商代晚期的东西，既没有遗迹，也没有遗物，这说明这群人是从别的地方迁徙到这里来的。从墓葬的形制、使用的葬具和腰坑、殉狗等现象看，这群人受中原文化的影响和渗透最迟在商代晚期已经发生了，但商代晚期这些人的祖先的居住地在哪里目前并不知道，从甲骨文和传世的文献记载我们知道，商代晚期山西有很多方国，而且有很多族群本身就属于戎狄，像鬼方可能就是媿姓狄人的祖先。鬼方是一个人群的集团，有很多分支，它们多居住在山区，也有一部分与中原华夏族群杂处在一个大区域中。《穆天子传》中记载穆天子（即周穆王）巡狩的时候曾经到过北方河宗的族群，见到郱伯，有人认为这个郱伯就是绛县横水墓地发现的倗伯，河宗氏当然距离黄河不会太远，河宗可能与鬼方有关，黄河中游两岸晋陕高原发现了很多商代晚期遗存，显示了黄河与这些文化遗存所属的人群之间密切的关系，若此推断可以成立的话，则大河口狄人族群与横水狄人族群的源头可能在晋陕高原黄河两岸，但他们何时迁到中原商文化区并不清楚。到西周时他们又被迁徙到了大河口与横水一带，在这里建国，并遗留下了文化遗迹和遗物。到了春秋早期以后大河口墓地就废弃了，生活遗址也不再使用了，说明西周时期以来的大河口人群又被迁徙到了别的地方，这个变故我们推测与晋献公时期的大肆扩张和兼并活动有关，倗国和霸国都是在这个时期被吞并的，后来他们被迁到了哪里目前也不清楚。虽然在大河口墓地周围也有东周遗址，但缺乏春秋中期的遗迹、遗物和墓葬，而且在大河口墓地发现几十座战国时期的窖穴破坏了西周墓葬，这充分说明春秋早期以后居住在大河口一带的人群与霸国的人群不是同一群人，他们各有自己的祖先与文化。

通过对大河口墓地人骨架的体质人类学特征研究，发现他们与曲村墓地的人骨具有更多的相似特征，绛县横水西周墓地的体质人类学研究也表明，媿姓狄人与中原华夏族群的相似度极高。这说明这些人群之间的差异更多的是文化上的差异，而不是生物特征上的差异，也从另外一个角度说明，几百万年以来人类在繁衍生息的过程中不断交流融合，在一定的自然历史区域中人群之间的生物性差异越来越小。

呦呦鹿鸣

燕国公主眼里的霸国

Harmonious Life: The State
of Ba in the Eyes of a Yan Princess

"人无礼则不生，事无礼则不成，国家无礼则不宁。"礼，作为"周人为政之精髓"，在西周时形成比较完善的制度，被孔子赞为"郁郁乎文哉"。据《礼记》记载，礼以婚礼为根本，丧礼、祭礼最为隆重，而和谐融洽体现于宴礼。这些日常礼仪，是霸国人处世的根本行为准则，政治理想和伦理道德都被规范其中，使人在这些仪式中接受礼的熏陶，从而使社会和谐，如呦呦鹿鸣般祥和、友睦。

霸国作为西周的封国，在史料中缺乏记载，随着霸国墓葬的发现，这个失落之国为世人所知晓。公元前10世纪的某天，燕国公主嫁到霸国，霸国的故事也由此展开。

Preface

As Xunzi (313BC—238BC), a great thinker and educator in the late Warring States period, once said "A man without rites is half man; a thing without rites is nothing; a country without rites is troublous." The word "rites" is what "*li*" means in the Confucian doctrine. "*Li*", or "rites", or "the ritual system", was highly praised by Confucius and had maturely developed during the Western Zhou Dynasty and became the norm of behavior. According to the *Book of Rites*, the wedding ceremony or "rite" was the most fundamental rite among other ritual ceremonies at the time. But in fact, the funeral and sacrificial ceremonies were more solemn in honoring the deceased, and a cozy and harmonious atmosphere was usually created to make the banquet go smoothly. As a vassal state of Zhou Dynasty, the Ba people considered the ritual norms as the moral principles in their daily life. They naturally accepted those ideas from the ordinary ceremonies to standardize the political ideals or social ethics. Thus the society became more harmonious and amicable, just like the peaceful and pleasant cry of the deer to one another.

Surprisingly, the State of Ba could not be found in any historical records until the discovery of the tombs of Ba in Shanxi Province. From that time on, a lost kingdom has come to light. In the 10th century B.C., a princess of the State of Yan married to a noble of Ba. Thus, we could know what had happened to her.

《诗经·郑风·女曰鸡鸣》

知子之来之，杂佩以赠之！
知子之顺之，杂佩以问之！
知子之好之，杂佩以报之！

Marriage:
Consolidate the
relationship between
Two Families

出嫁

合二姓之好

两姓同婚涉及两姓联姻的质量和稳定性，关系到宗族是否昌盛。《礼记》中明确记载："昏（同婚）礼者，将合二姓之好，上以事宗庙，而下以继后世也，故君子重之"。与燕国公主有关的器物埋葬在数百公里之外的山西霸国，最可能的原因是她嫁给了霸国的某位君主。燕国公主的出嫁，便蕴含着两个宗族之间诸多美好的期望。

In accordance with the *Record of Rites*, the purpose of traditional marital union between two families was to carry on the family line, so the spirits of the deceased ancestors could be honored. It may be quite true that a successful marriage could not only help keep the family clans thriving and prosperous but also secure the family members blessed and favorable in the future. The reason why the bronze wares owned by the princess of Yan were discovered in Shanxi is that presumably she might have been married to a prince of Ba State. Thus, we assume that the two noble family clans probably expected to be benefited in the form of marriage connection and adding glory to their names.

揭 秘 霸 国

Discovery of State of Ba

2007 年 5 月，在山西省临汾市翼城县大河口，一个墓葬群的发现，轰动了考古界。史籍中没有丝毫印迹的西周封国——霸国随之横空出世，其数量惊人的随葬品、风格独特的葬俗、精美的漆木器、目前中原地区发现最早的漆木人俑、填补史籍记载空白的青铜器铭文、保存完好的原始瓷器……众多令人惊叹的发现究竟该作何解释？好奇的人们无不翘首期待着答案。

令人惊喜的是，在一座大墓中发现的一件青铜器上有铭文"燕侯旨作姑妹宝尊彝"，表明燕侯旨小姑姑的器物在山西被发现。燕国公主嫁给了谁？燕国公主的婆家——霸国，为什么史料中不见记载？青铜器铭文为我们娓娓道出了霸国君主霸伯及其诸多的秘密。

大河口墓地位于翼城县隆化镇大河口村北台地上，西距翼城县城约 6 公里，墓地北高南低，北倚二峰山，南面浍河支流，西北为浍河干流。其所处的山西省南部，是一处"藏龙卧虎"的地方，仅西周时期的墓葬和遗址就有数百处之多。

历史文献曾多次记载，晋国的开国始祖唐叔虞被封在了山西翼城一带，特别是今翼城县有很多与"唐"有关的地名，如北唐村、南唐村、东唐村等。因此，在大河口墓地发掘之前，考古人员更倾向于认为这个地方是"唐"的墓地。但随着大河口墓地考古工作的进展，大量带"霸"

霸国档案									
中文名称	英文名称	国君	国姓	政治体制	官方语言	居住地域	存在时间		
霸国	State of Ba	霸伯	媿（音愧）姓狄人	君主制	上古汉语	山西省临汾市翼城县	西周（公元前 1046 年—公元前 771 年）		

大河口墓地位置示意图

字铭文的青铜器以无可辩驳的事实证明了一个史书阙载的古国——霸国的存在，随之出土的丰富文物展示出霸国独具特色的文化。

大河口墓地南北长约 300 米，东西宽约 150 米，面积约 4.5 万平方米，墓葬 1000 余座，时代自西周早期延续至两周之际。

墓地系由 2007 年 5 月遭盗掘而被首次发现，其中以霸伯墓葬最为引人注目。2008 年 9 月至 2011 年 5 月，山西省考古研究所进行了全面勘探和大面积发掘，共发掘 1.6 万平方米，发现近 600 座墓葬，总计出土文物 15000 余件套，其中青铜容器约 220 件，锡器 50 余件，陶器 600 余件。目前，第三阶段的考古发掘已经开始。

大河口墓地，是继曲村墓地、横水墓地之后，西周考古学史上又一次重大发现，获得了"2010 年度全国十大考古新发现""2009—2010 年度国家文物局田野考古一等奖"等荣誉。

大河口墓地航拍图（自东向西拍摄）

"霸中"簋

Gui food container with the inscription
"Ba Zhong"

西周（公元前 1046 年 – 公元前 771 年）
高 25 厘米，耳间距 28 厘米，口径 18.7 厘米
山西省临汾市翼城县大河口墓地出土
山西省考古研究所藏

Western Zhou Dynasty (1046 B.C. –771 B.C.)
Height: 25 cm, distance between ears: 28 cm, mouth
diameter: 18.7 cm
Unearthed from Dahekou tomb complex in Yicheng
County, Linfen, Shanxi Province
Shanxi Provincial Institute of Archaeology

　　器内底和盖内均有铭文"霸中作旅彝"。
"霸"就是霸伯的国族名。"霸伯"为霸国的
国君。这里的"伯"应为嫡庶排位，而非爵位。
这件青铜簋有"霸中"的铭文，"中"就是"仲"，
在兄弟排行中代表第二，属次子。

在文献中找不到霸国的记载，一种原因可能是关于霸国的记载在漫长的历史流传中遗失了；另一种原因是，霸国在当时是个很小的国家，人口不多，其所居城邑和所辖区域也不会很大，而在西周时期，像霸国这样的小国数量众多，所以被传统的史料遗漏了。

带"霸"字铭文的青铜器曾见于古文字著作中，如《殷周金文集成》中著录有"霸姞作宝尊彝"鼎和簋，同时在已发掘的墓地中也有出现，如在山西省曲沃县西周晋国曲村墓地就出土了一件"霸伯作宝尊彝"铜簋。这意味着，大河口墓地为之前在其他各处发现的"霸"器找到了真正的归宿。

释文：
霸姞作
宝尊彝

霸姞鼎铭文拓片

（《殷周金文集成》4.2184.2）

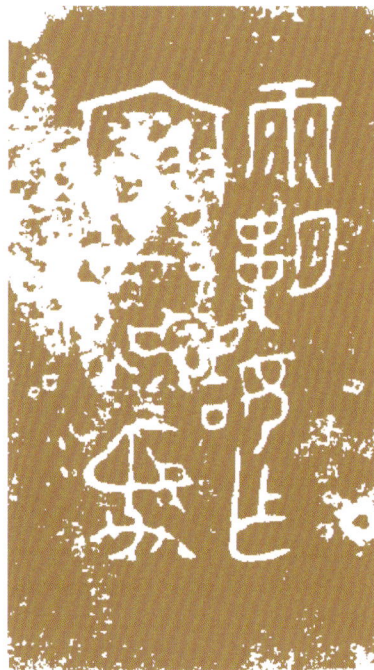

释文：
霸姞作
宝尊彝

霸姞簋铭文拓片

（《殷周金文集成》6.3565）

铜簋

Gui food container

西周（公元前1046年－公元前771年）

高13.4厘米，口径16.9厘米

山西省曲沃县曲村6197号墓出土

山西博物院藏

Western Zhou Dynasty (1046 B.C. –771 B.C.)
Height: 13.4 cm, mouth diameter: 16.9 cm
Unearthed from tomb No.6197 in Qu Village,
Quwo County, Shanxi Province
Shanxi Museum

器内底有铭文"霸伯作宝尊彝"。

释文：
霸伯作
宝尊彝

铜簋铭文拓片

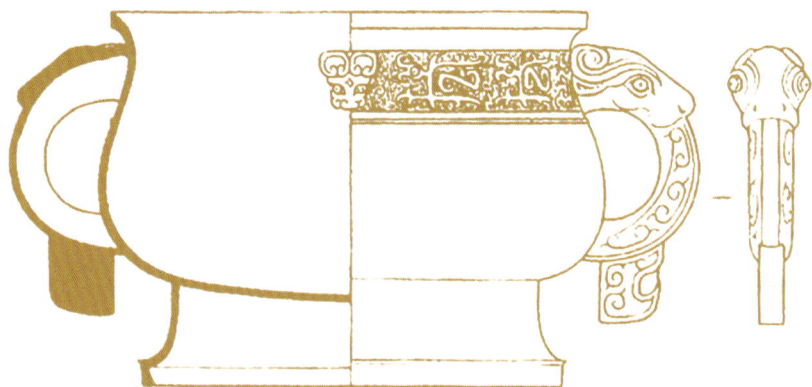

霸伯罍

Lei wine container with the inscription "Ba Bo"

西周（公元前 1046 年 – 公元前 771 年）

高 34.5 厘米，口径 15.5 厘米，耳间距 32 厘米

山西省临汾市翼城县大河口墓地出土

山西省考古研究所藏

Western Zhou Dynasty (1046 B.C. –771 B.C.)
Height: 34.5 cm, mouth diameter: 15.5 cm, distance between ears: 32 cm
Unearthed from Dahekou tomb complex in Yicheng County, Linfen, Shanxi Province
Shanxi Provincial Institute of Archaeology

盛酒器。器口内壁有铭文"霸伯作宝尊"。

伯方鼎

Square *ding* cauldron with the inscription "Bo Zuo Ding"

西周（公元前 1046 年 - 公元前 771 年）
高 19.6 厘米，耳间距 16.7 厘米
山西省临汾市翼城县大河口墓地出土
山西省考古研究所藏

Western Zhou Dynasty (1046 B.C. –771 B.C.)
Height: 19.6 cm, distance between ears: 16.7 cm
Unearthed from Dahekou tomb complex in
Yicheng County, Linfen, Shanxi Province
Shanxi Provincial Institute of Archaeology

器内壁有铭文"伯作鼎"。

伯方鼎出土照片

兽面纹三足圆鼎

Ding tripod cauldron with beast face design

西周（公元前 1046 年－公元前 771 年）

高 55 厘米，耳间距 42 厘米

山西省临汾市翼城县大河口墓地出土

山西省考古研究所藏

Western Zhou Dynasty (1046 B.C. –771 B.C.)
Height: 55 cm, distance between ears: 42 cm
Unearthed from Dahekou tomb complex in
Yicheng County, Linfen, Shanxi Province
Shanxi Provincial Institute of Archaeology

夔凤形足鼎

Ding tripod cauldron with
phoenix-shaped legs

西周（公元前 1046 年 – 公元前 771 年）
高 20.3 厘米，耳间距 14 厘米
山西省临汾市翼城县大河口墓地出土
山西省考古研究所藏

Western Zhou Dynasty (1046 B.C. –771 B.C.)
Height: 20.3 cm, distance between ears: 14 cm
Unearthed from Dahekou tomb complex in
Yicheng County, Linfen, Shanxi Province
Shanxi Provincial Institute of Archaeology

涡纹圈足簋

Gui food container with paisley pattern

西周（公元前 1046 年－公元前 771 年）

高 18.3 厘米，耳间距 35.5 厘米，圈足径 19.1 厘米

山西省临汾市翼城县大河口墓地出土

山西省考古研究所藏

Western Zhou Dynasty (1046 B.C. –771 B.C.)
Height: 18.3 cm, distance between ears: 35.5 cm,
base diameter: 19.1 cm
Unearthed from Dahekou tomb complex in
Yicheng County, Linfen, Shanxi Province
Shanxi Provincial Institute of Archaeology

在同时期一定地域范围内，墓葬的方向与墓主人的族属有着很大的关系，这是人群融合过程中最容易保留和传承下来的习俗，也是最不容易被改变的习俗。以往在晋南地区发现的西周墓葬大多为南北向，特别是曲沃县北赵村晋侯墓地的晋侯夫妇墓葬都是南北向的。这与霸国墓地、绛县横水倗国墓地发现的墓葬截然不同。大河口墓地头向西的墓葬 540 余座，头向东的墓葬 30 余座，南北向墓葬仅 4 座。

山西西周墓葬分析表

墓葬名称	主要头向	斜洞	腰坑	祭狗	殉人	族属
北赵晋侯墓地	向北	无	无	大量	一例	周人
天马—曲村墓地	以北、东为主，有少量头向西的墓葬	无	很少	有一些	极少	头向北的墓主人可能是以周人为主体的族群，头向东的墓主人可能是以唐人为主体的族群，头向西的墓主人可能是以"怀姓九宗"为主体的族群
倗国墓地	以向西为主	有	较多	较多	较多	媿姓狄人
霸国墓地	以向西为主	有	较多	较多	无	媿姓狄人

注：①倗国，史书无记载，位于山西省绛县横水镇横北村的一个西周诸侯国，发现于 2004 年；
②北赵晋侯墓地和曲村墓地是西周时期的晋国墓地，位于山西省曲沃县。

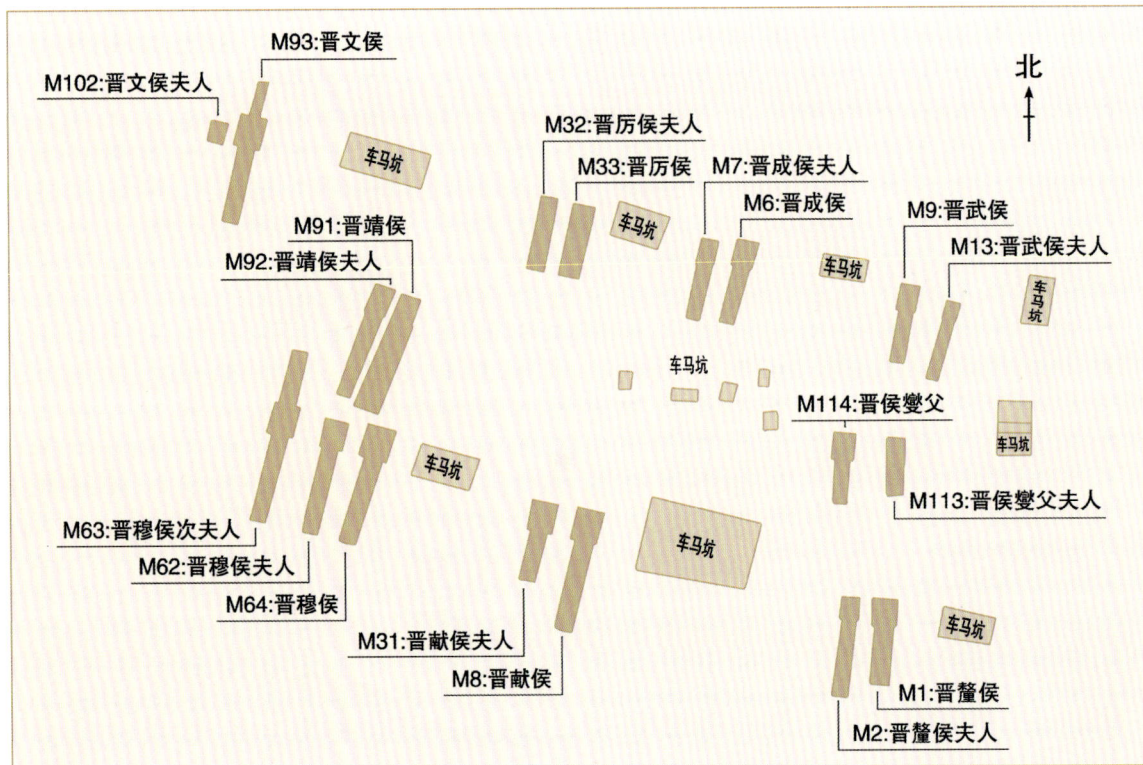

北赵晋侯墓地平面图

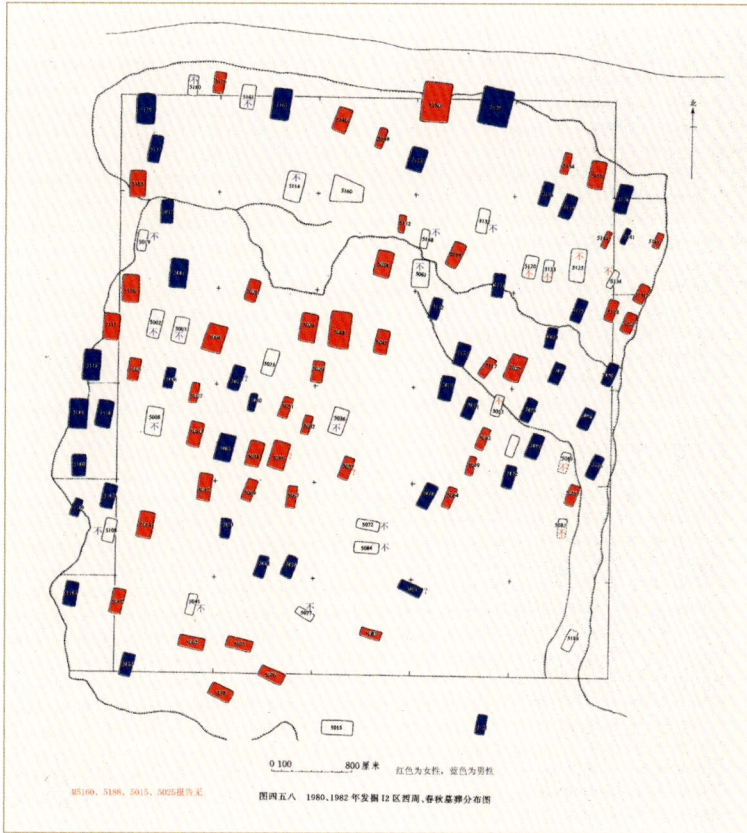

曲村墓地Ⅰ2区平面图

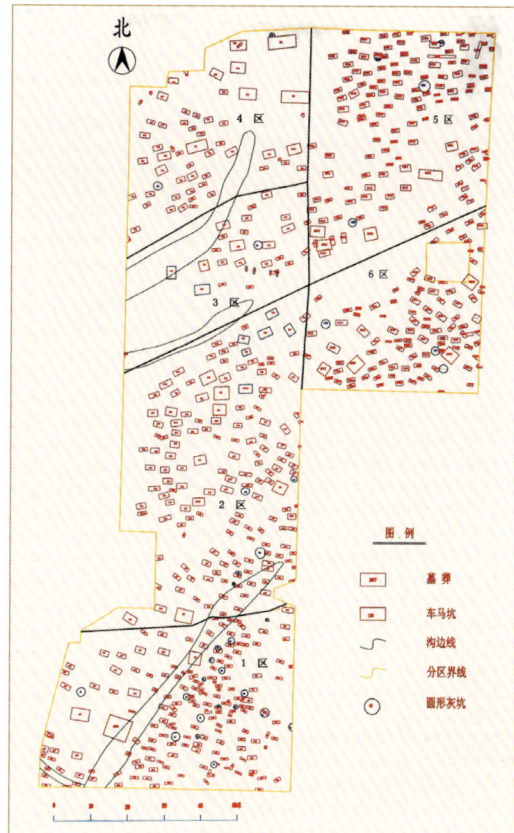

翼城县大河口墓地平面图

倗国墓地航拍图

大河口一座墓葬的腰坑。一般腰坑中都有一只小狗的骨架，当是埋葬前的祭祀用牲

横水倗国墓地 M2036 腰坑

倗仲鼎铭文拓片

释文：
倗仲作毕媿媵鼎，其
万年宝用

其铭意为倗仲将女儿媿氏嫁于毕国君主为妃，为此铸造了此鼎，愿其万年永宝此物。

商周时期，人们在入葬之前，都会举行一个祭奠仪式，于墓主人身下的腰坑（实际上就是祭奠坑）里祭狗就是其中的一种形式。从中原地区来看，腰坑是商文化的一种习俗。商灭亡后，原先受商文化影响的人，包括商遗民以及商统治过的邦国百姓，仍一直延续着这种习俗。此外，霸国墓地还通常用大狗来殉葬，这很有可能与墓主人生前饲养习惯和埋葬习俗有关，殉大狗或许可起到守护、驱鬼等作用。

据《通志·氏族略序》记载，"贵者有氏，贱者有名无氏"，"姓所以别婚姻，氏所以别贵贱"。姓作为族号，代表了共同的血缘和血统，不同的族群有着不同的文化和习俗。那么，霸国的姓氏又是什么呢？

一方面，从古文字学来看，在上古音里 b 和 p 不分，因此在古代，"霸"字既可读作 bà 亦可读作 pò，这就意味着霸国在过去或许也可称为 pò 国，而同音字"魄"字的右半边是个"鬼"字，所以，"魄"字和鬼有关，而鬼又与鬼方有关。鬼方是夏商时期的方国群落，属于戎狄族群。东周文献中有明确记载，"鬼方"就是媿（音愧）姓狄人的前身。

另一方面，从考古学文化因素来看，根据传世的倗仲鼎铭文"倗仲作毕媿媵鼎"可知倗为媿姓，媿姓在文献中被记载为东周时赤狄的姓，那么可以推测倗国墓地的族群为狄人。考古发掘资料显示，大河口墓地与横水墓地在墓主人头向、斜洞、腰坑和祭狗等方面有着十分明显的共性。所以，大河口墓地可能也是媿姓狄人的墓地，它与横水墓地的人群是狄人的不同分支。通过上述分析，我们可以揭晓燕国公主嫁给了媿姓狄人。

铜 鬲

Li tripod steamer

西周（公元前 1046 年 – 公元前 771 年）
高 16.8 厘米，耳间距 12.6 厘米
山西省临汾市翼城县大河口墓地出土
山西省考古研究所藏

Western Zhou Dynasty (1046 B.C. –771 B.C.)
Height: 16.8 cm, distance between ears: 12.6 cm
Unearthed from Dahekou tomb complex in Yicheng County,
Linfen, Shanxi Province
Shanxi Provincial Institute of Archaeology

　　炊器或食器。口颈内壁有铭文"作父癸鬲"，意思就是给其父亲癸做的这件鬲。"癸"是日名，是以十天干起的名字。研究者认为日名多是商代的习俗，商人以日为神。曾有人提出周人不用日名之说，但燕国使用日名，燕侯旨的器物上就有日名；大河口墓葬中发现的青铜爵上也有父辛的日名，至少在姬姓周人的燕国这种情况与所谓的周人不用日名说相悖；在河南平顶山市发现的应国墓地的青铜器铭文上也有日名，应国的应侯也是姬姓。但晋国的上层统治者姬姓贵族就没有使用日名的习俗。所以，大部分周人不用日名的观点才可能是接近历史真实的。

铜 豆

Dou food container

西周（公元前 1046 年－公元前 771 年）
高 21.8 厘米，口径 16.4 厘米，足径 15.4 厘米
山西省临汾市翼城县大河口墓地出土
山西省考古研究所藏

Western Zhou Dynasty (1046 B.C. –771 B.C.)
Height: 21.8 cm, mouth diameter: 16.4 cm, base
diameter: 15.4 cm
Unearthed from Dahekou tomb complex in Yicheng
County, Linfen, Shanxi Province
Shanxi Provincial Institute of Archaeology

内壁底上有"霸伯作太庙宝尊彝……"铭文。

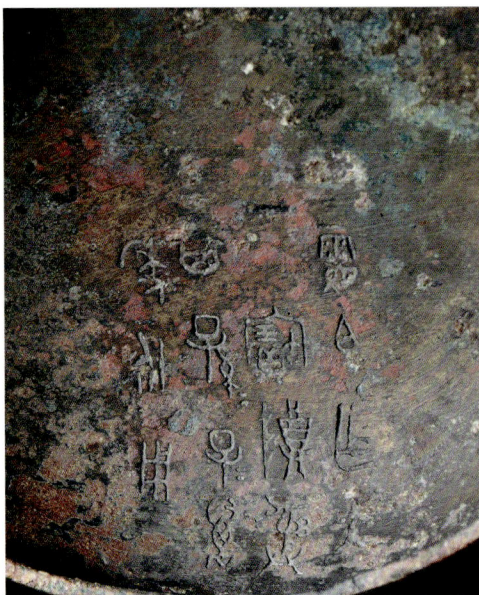

铜豆出土现场照片

西周时期，周王朝分封了大量的诸侯。为了维持长治久安的局面，各诸侯国除了壮大自己的国力之外，处理好与周王朝和周边国家的关系也至关重要。诸侯之间一般通过盟会、相互聘问联络感情。在霸国所处的山西晋南地区，西周诸侯国有二十余个。在众多诸侯国中，霸国处于怎样的地位？它与周王朝和周边邻居的关系如何？它用什么方法来进行文化交流？关于上述种种疑问，霸国墓葬群出土的文物透露出了蛛丝马迹。

各诸侯国位置示意图

扁腹簋

　　这件扁腹簋的盖内壁和器底都有较长的铭文，大意为十一月井叔代表周王拜会霸伯。为嘉奖、勉励霸伯在征伐方面的成绩，赐予他蒙车毂的皮革一百、丹砂二丼、嬮地所产的虎皮一张。霸伯拜手叩头，报答颂扬井叔的美意。因此做了宝簋，愿其子子孙孙永宝此物。

三足簋

Gui tripod food container

西周（公元前 1046 年 – 公元前 771 年）
高 22.8 厘米，耳间距 25.3 厘米
山西省临汾市翼城县大河口墓地出土
山西省考古研究所藏

Western Zhou Dynasty (1046 B.C. –771 B.C.)
Height: 22.8 cm, distance between ears: 25.3 cm
Unearthed from Dahekou tomb complex in
Yicheng County, Linfen, Shanxi Province
Shanxi Provincial Institute of Archaeology

　　三足簋盖内壁和器底有铭文"芮公
舍霸马两玉金用铸簋"。"舍"就是赐
予的意思，大意是说周王室重臣芮公赐
给霸伯两匹马、玉和青铜，用来铸造这
件簋。

尚 盂

Yu vessel

西周（公元前 1046 年 – 公元前 771 年）

高 34.2 厘米，耳间距 42.5 厘米，口径 39.5 厘米

山西省临汾市翼城县大河口墓地出土

山西省考古研究所藏

Western Zhou Dynasty (1046 B.C. –771 B.C.)
Height: 34.2 cm, distance between ears: 42.5 cm, mouth diameter: 39.5 cm
Unearthed from Dahekou tomb complex in Yicheng County, Linfen, Shanxi Province
Shanxi Provincial Institute of Archaeology

　　盛水器或盛食器。盂内壁有一篇反映西周聘礼的长篇铭文，共 116 字。其中有"唯三月，王使伯考蔑尚历……霸伯拜、稽首，对扬王休，用作宝盂，孙子子其万年永宝"，意思是伯考代表周王来霸国赏赐和勉励霸伯，霸伯回赠伯考和周王礼物。这篇铭文可以与《仪礼·聘礼》互相印证。

铜人顶盘（灯）

Oil lamp with human-shaped support

西周（公元前 1046 年 – 公元前 771 年）
高 13.2 厘米，盘耳间距 7.2 厘米，盘口径 6.8 厘米
山西省临汾市翼城县大河口墓地出土
山西省考古研究所藏

Western Zhou Dynasty (1046 B.C. –771 B.C.)
Height: 13.2 cm, distance between handles: 7.2
cm, mouth diameter: 6.8 cm
Unearthed from Dahekou tomb complex in
Yicheng County, Linfen, Shanxi Province
Shanxi Provincial Institute of Archaeology

　　这件铜人形灯，与战国时期的豆形灯形态如出一辙，应当是战国灯具的源头。据考古资料看，此灯很可能是中国目前发现最早的青铜灯。

　　大河口墓地出土的铜人顶盘与传世的晋侯铜人形制极其相似。晋侯铜人敝膝上有一篇 21 个字的铭文："唯五月，淮夷伐格，晋侯搏戎，获厥君冢师，侯扬王于兹。"大意是：在五月的一天，南边的淮夷戎族部落来讨伐格国（即霸国），晋侯与淮夷进行了搏战以救援霸国，俘获了淮夷的君主，晋侯为颂扬周王的美德铸作此器。霸国当时应该是一个小国，晋国是大国，可能是晋南地区的方伯，在向周王室纳贡和对抗外敌入侵方面具有组织救助的职责。这篇铭文为研究晋与格（霸）的关系提供了非常重要的资料。

晋侯铜人

陶三足盘

Pottery food container

西周（公元前 1046 年－公元前 771 年）

高 9.5 厘米

山西省绛县横水倗国墓地 M2113 出土

山西省考古研究所藏

Western Zhou Dynasty (1046 B.C. –771 B.C.)
Height: 9.5 cm
Unearthed from M2113 of Peng State in Hengshui, Jiang County, Shanxi Province
Shanxi Provincial Institute of Archaeology

陶三足盘

Pottery food container

西周（公元前 1046 年－公元前 771 年）

高 9 厘米，直径 17.8 厘米

山西省临汾市翼城县大河口墓地出土

山西省考古研究所藏

Western Zhou Dynasty (1046 B.C. –771 B.C.)
Height: 9 cm, diameter: 17.8 cm
Unearthed from Dahekou tomb complex in Yicheng County, Linfen, Shanxi Province
Shanxi Provincial Institute of Archaeology

故宫博物院藏有一件佣生簋（又名格伯簋），属西周共王时代，器内底有铭文八行八十二字，大意是格伯用良马四匹换取佣生三十田，双方分执契券，勘定田界。在古文字中，"霸"与"格"为同字异构，二字古音相通，写法相似或相同。佣生就是佣国外甥的意思，这件器物铭文显示出霸国与佣国之间的交往。

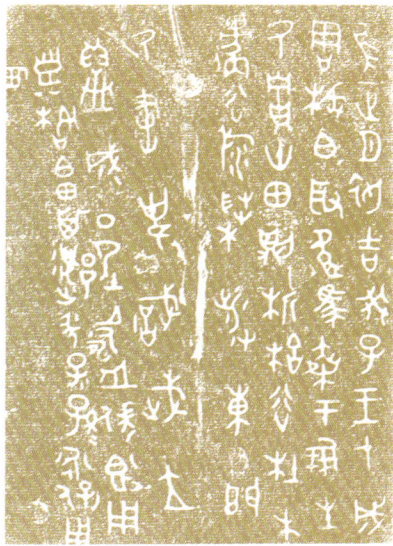

佣生簋铭文拓片

公 主 出 阁

Long Distance Marriage

《礼记》中说："昏（同婚）礼，万世之始也。取于异姓，所以附远厚别也"，婚礼意味着以后万世子孙都从此时开始，意义深远。迎娶异姓女子为妻，一是为了联合疏远的异姓成为姻亲，一是实行同姓不婚的需要。同姓不婚的一个直接后果就是女子远嫁。燕国公主是燕侯旨的小姑姑，为庆贺她嫁到霸国，燕侯旨特意铸造青铜器作为礼物。"玉骨久沉泉下土"，与燕国公主有关的器物在霸国墓葬被发现，揭开了燕霸两国联姻的历史。

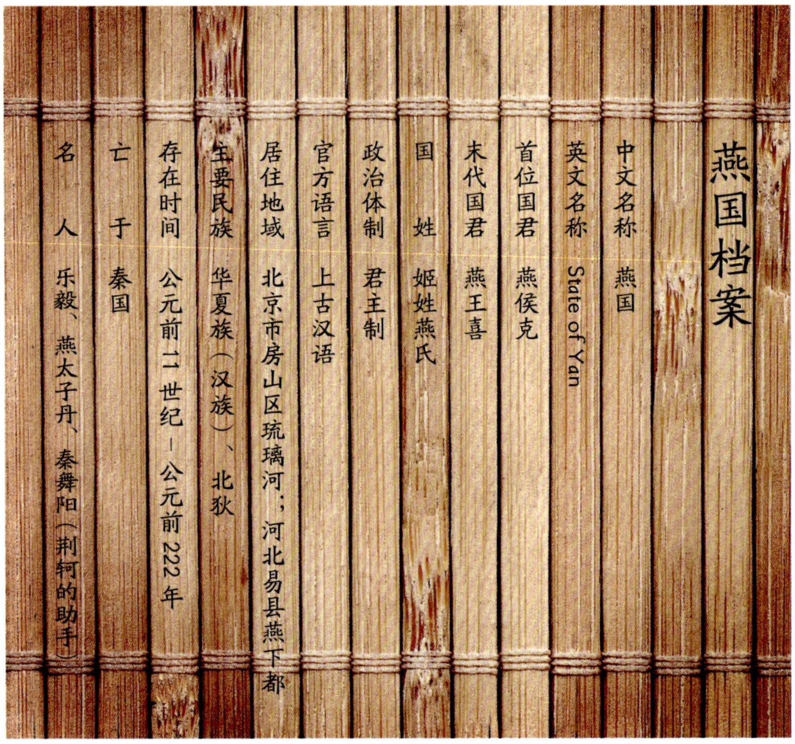

燕国档案

中文名称　燕国

英文名称　State of Yan

首位国君　燕侯克

末代国君　燕王喜

国　姓　姬姓燕氏

政治体制　君主制

官方语言　上古汉语

居住地域　北京市房山区琉璃河；河北易县燕下都

主要民族　华夏族（汉族）、北狄

存在时间　公元前二世纪 — 公元前222年

亡　于　秦国

名　人　乐毅、燕太子丹、秦舞阳（荆轲的助手）

"公主"这个称呼，在春秋战国时期才出现，更早的时候，她们被称为"王姬"。在周朝的时候，"姬"本是周天子的姓氏，后来也作为形容妇人的美称。人们认为这个字比较高贵，而周天子的女儿也比其他王族的女儿要高贵，能够配得上这个"姬"字，所以就叫"王姬"。

　　"王姬"结婚要有主婚人。一般来说，家长要做主婚人，可"王姬"的家长是周天子，地位非常尊贵，无法做主婚人。同时主婚人又要负责很多杂事，古人就在比天子低一级的诸侯里挑人选。由于要主婚，所以这个人选一定要和天子同姓。而那个时候的诸侯，级别最高的就叫做"公"，因此"王姬"的定义就变成了需要由"公"来主持其婚事的女子。所以，她们又被称为"公主"*了。

　　此外，还有一种解释：《春秋指掌碎玉》中说，所谓"公"，指的是"三公"。在周代，"三公"是指太师、太傅、太保。

貘耳提梁卣

You wine container with overtop handle and tapir-head-shaped ears

西周（公元前 1046 年 - 公元前 771 年）
高 23 厘米，耳间距 26.5 厘米
山西省临汾市翼城县大河口墓地出土
山西省考古研究所藏

Western Zhou Dynasty (1046 B.C. –771 B.C.)
Height: 23 cm, distance between ears: 26.5 cm
Unearthed from Dahekou tomb complex in
Yicheng County, Linfen, Shanxi Province
Shanxi Provincial Institute of Archaeology

　　貘，音莫。这件貘耳提梁卣的盖子尚未打开，通过 X 射线扫描发现里面有两件觯，另外至少还有四件器物，器形不明。

出嫁
合二姓之好

*为了方便于读者理解，本书中的燕国王姬权且称为"公主"。

"燕侯旨"卣

You wine container with the inscription "Yan Hou Zhi"

西周（公元前 1046 年 - 公元前 771 年）
高 34.5 厘米，两耳间距 29 厘米
山西省临汾市翼城县大河口墓地出土
山西省考古研究所藏

Western Zhou Dynasty (1046 B.C. –771 B.C.)
Height: 34.5 cm, distance between ears: 29 cm
Unearthed from Dahekou tomb complex in
Yicheng County, Linfen, Shanxi Province
Shanxi Provincial Institute of Archaeology

"燕侯旨"卣铭文

卣，音酉。盛酒器。在这件带盖青铜卣内，放置着一套酒器，共七件，分别是斗一件、单耳罐一件、大小不同的觯五件。青铜卣的盖内面和器底内面都铸有一篇铭文，内容是"燕侯旨作姑妹宝尊彝"，"姑妹"是小姑姑的意思。这是燕侯旨给他的小姑姑燕国公主制作的器物。

克盉

He wine container inscribed with Ke

西周早期（前11世纪中期－前10世纪中期）

通高 26.8 厘米，口径 14 厘米

北京市房山区琉璃河出土

首都博物馆藏

Early Western Zhou Dynasty (mid. 11th century B.C.–
mid. 10th century B.C.)
Height: 26.8 cm, mouth diameter: 14 cm
Unearthed from Liulihe, Fangshan District, Beijing
Capital Museum

　　铭文为：王曰："太保，隹乃明乃鬯，享
于乃辟，余大对乃享。"令克侯于匽，使羌、马、䚢、
于、驭、微。克宅匽，入土众有司，用作宝尊彝。

　　大意是：周王说，太保，你贤明畅达，我
非常赞赏你关于治国安邦的方法和观点。周王
册封太保的长子"克"为燕侯，管理羌、马、䚢、于、
驭、微等部族。克来到燕地，接收了土地和管
理机构，为了纪念此事制作了这件珍贵的礼器。
"克"是周初召公奭（音是）的儿子。

克盉铭文

克罍

Lei wine container inscribed with Ke

西周早期（公元前11世纪中期－
公元前10世纪中期）

通高 32.7 厘米，口径 14 厘米

北京市房山区琉璃河出土

首都博物馆藏

Early Western Zhou Dynasty (mid. 11th
century B.C. – mid. 10th century B.C.)
Height: 32.7 cm, mouth diameter: 14 cm
Unearthed from Liulihe, Fangshan
District, Beijing
Capital Museum

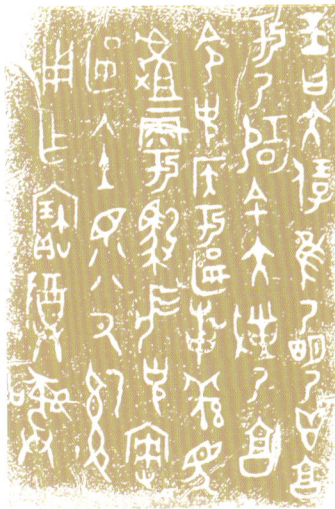

铭文为：王曰："太保，隹乃明乃鬯，
享于乃辟，余大对乃享。"令克侯于匽，使羌、
马、譶、于、驭、微。克宅匽，入土眾有司，
用作宝尊彝。

大意是：周王说，太保，你贤明畅达，
我非常赞赏你关于治国安邦的方法和观点。
周王册封太保的长子"克"为燕侯，管理羌、
马、譶、于、驭、微等部族。克来到燕地，
接收了土地和管理机构，为了纪念此事制
作了这件珍贵的礼器。

旨，是燕国第二代国君的名字。有"燕侯旨"铭文的器物此前仅见到两件，一件在日本，一件在上海博物馆，都是传世品。日本藏燕侯旨鼎的铭文为："燕侯旨初见事于宗周。王赏旨贝廿朋，用作又始宝尊彝。"大意为："燕侯旨第一次入朝觐见周王，周王赏赐了贝二十朋（串）。旨于是作了祭祀姒的彝器。""又始"读为"有姒"，根据周代同姓不婚的规定可以看出姬姓召公家族曾与姒姓通婚。

上海博物馆藏燕侯旨鼎铭文为："燕侯旨作父辛尊。"其铭文大意为："燕侯旨为其父亲所做的鼎"。据考证，燕侯旨应为召公奭的孙子，比克晚一辈，其父亲庙号为"父辛"。

日本藏燕侯旨鼎及铭文拓片

上海博物馆藏燕侯旨鼎及铭文拓片

出嫁 合二姓之好

61

铜"旨"爵

Jue drinking vessel with the
inscription "Zhi Zuo"

西周（公元前 1046 年 - 公元前 771 年）
高 23 厘米，流尾长 18 厘米
山西省临汾市翼城县大河口墓地出土
山西省考古研究所藏

Western Zhou Dynasty (1046 B.C. –771 B.C.)
Height: 23 cm, length of spout: 18 cm
Unearthed from Dahekou tomb complex in
Yicheng County, Linfen, Shanxi Province
Shanxi Provincial Institute of Archaeology

温酒器或饮酒器。柱面有铭文"旨
作"，鋬后腹部有铭文"父辛爵世"。

铜"旨"爵

Jue drinking vessel with the inscription "Zhi Zuo"

西周（公元前 1046 年 – 公元前 771 年）

残高 23.2 厘米，柱间距 7.2 厘米，流尾长 17 厘米，
足高 7.5 厘米

山西省临汾市翼城县大河口墓地出土

山西省考古研究所藏

Western Zhou Dynasty (1046 B.C. –771 B.C.)
Remained height: 23.2 cm, distance between legs: 7.2 cm,
length of spout: 17 cm, height of leg: 7.5 cm
Unearthed from Dahekou tomb complex in Yicheng County,
Linfen, Shanxi Province
Shanxi Provincial Institute of Archaeology

　　其中一柱面有铭文"旨作"，鋬后腹部有铭文"父
辛爵世"。

《诗经·召南·鹊巢》

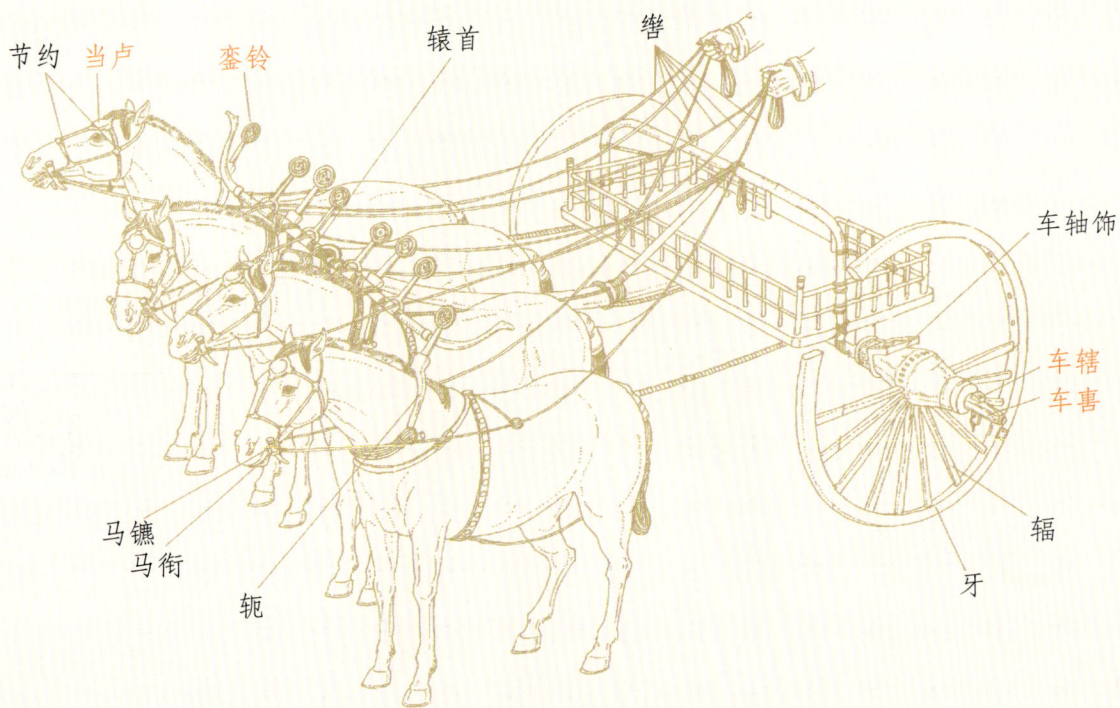

维鹊有巢，维鸠居之。
之子于归，百辆御之。
维鹊有巢，维鸠方之。
之子于归，百辆将之。
维鹊有巢，维鸠盈之。
之子于归，百辆成之。

节约　当卢　銮铃　　辕首　　辔

车轴饰

车辖
车軎

辐

牙

马镳
马衔

轭

车马器名称示意图

大河口墓地铜銮铃出土照片

铜车軎、车辖

Locking hub to the axle and linchpin

西周（公元前 1046 年－公元前 771 年）

车軎通高 12.5 厘米，顶径 4.8 厘米，底径 5.6 厘米

车辖通长 11.4 厘米，宽 3.5 厘米

山西省临汾市翼城县大河口墓地出土

山西省考古研究所藏

Western Zhou Dynasty (1046 B.C. –771 B.C.)
Axle: full height: 12.5 cm, top diameter: 4.8 cm,
bottom diameter: 5.6 cm
Linchpin: full length: 11.4 cm, width: 3.5 cm
Unearthed from Dahekou tomb complex in Yicheng
County, Linfen, Shanxi Province
Shanxi Provincial Institute of Archaeology

铜车軎、车辖

Locking hub to the axle and linchpin

西周（公元前 1046 年－公元前 771 年）

车軎通高 12.5 厘米，顶径 4.4 厘米，底径 5.6 厘米

车辖通长 11.4 厘米，宽 3.5 厘米

山西省临汾市翼城县大河口墓地出土

山西省考古研究所藏

Western Zhou Dynasty (1046 B.C. –771 B.C.)
Axle: full height: 12.5 cm, top diameter: 4.4 cm, bottom
diameter: 5.6 cm
Linchpin: full length: 11.4 cm, width: 3.5 cm
Unearthed from Dahekou tomb complex in Yicheng County,
Linfen, Shanxi Province
Shanxi Provincial Institute of Archaeology

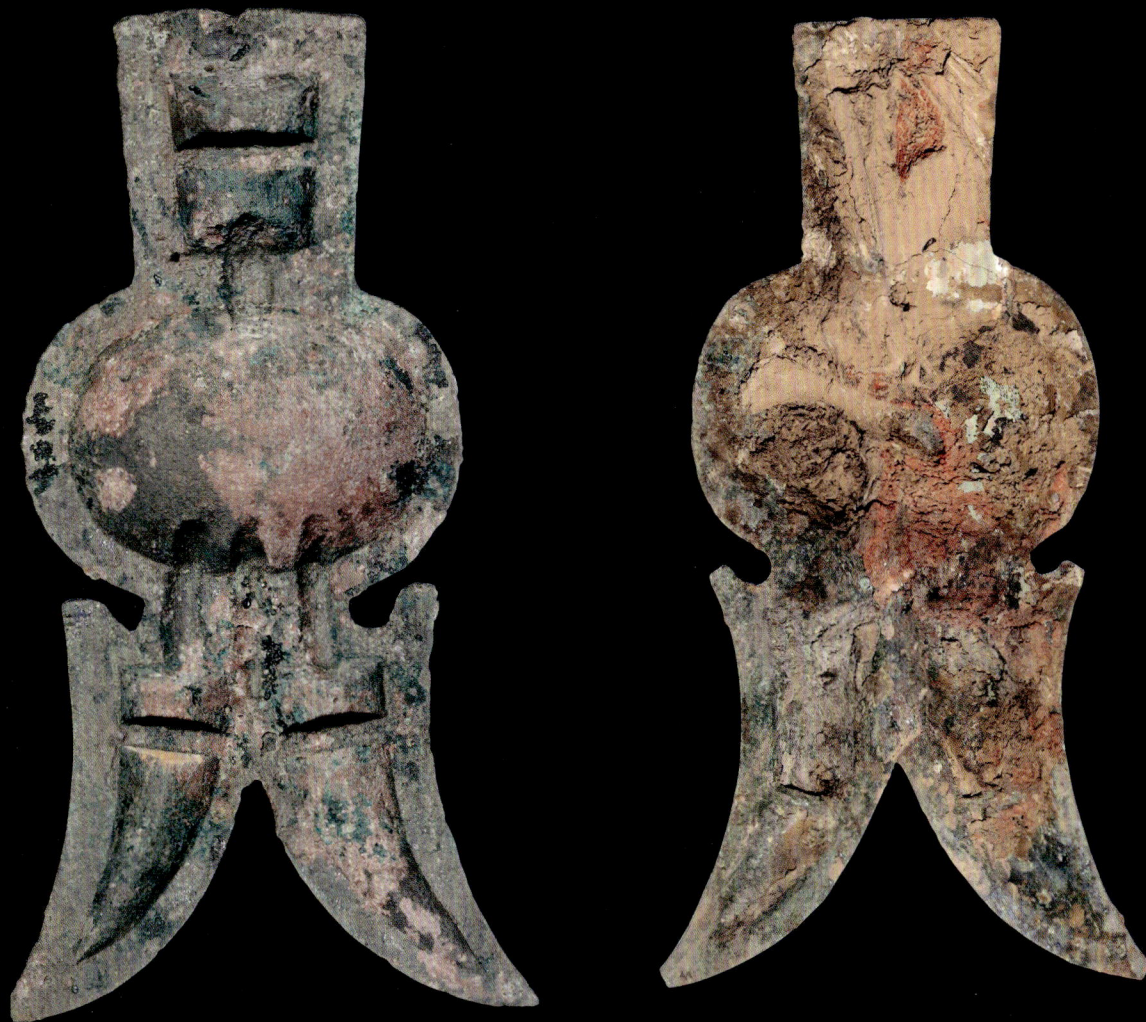

铜当卢

Ornaments of bridle

西周（公元前 1046 年－公元前 771 年）

左：高 19.5 厘米，最宽 10.2 厘米，鼻梁宽 4.7 厘米

右：高 18.7 厘米，最宽 9.5 厘米，鼻梁宽 4.2 厘米

山西省临汾市翼城县大河口墓地出土

山西省考古研究所藏

Western Zhou Dynasty (1046 B.C. –771 B.C.)
Left: Height: 19.5 cm, largest width: 10.2 cm, width of handle:
4.7 cm
Right: Height: 18.7 cm, largest width: 9.5 cm, width of handle:
4.2 cm
Unearthed from Dahekou tomb complex in Yicheng County,
Linfen, Shanxi Province
Shanxi Provincial Institute of Archaeology

铜銮铃

Luan bell

西周（公元前 1046 年－公元前 771 年）

高 15.3 厘米，铃长径 8.8 厘米，底座 4.4 厘米 × 3.1 厘米

山西省临汾市翼城县大河口墓地出土

山西省考古研究所藏

Western Zhou Dynasty (1046 B.C. −771 B.C.)
Height: 15.3 cm, largest diameter: 8.8 cm, base: 4.4 cm×3.1 cm
Unearthed from Dahekou tomb complex in Yicheng County, Linfen,
Shanxi Province
Shanxi Provincial Institute of Archaeology

铜銮铃

Luan bell

西周（公元前 1046 年－公元前 771 年）

高 15 厘米，铃长径 8.8 厘米，底座 4.3 厘米 × 3 厘米

山西省临汾市翼城县大河口墓地出土

山西省考古研究所藏

Western Zhou Dynasty (1046 B.C. −771 B.C.)
Height: 15 cm, largest diameter: 8.8 cm, base: 4.3 cm×3 cm
Unearthed from Dahekou tomb complex in Yicheng County,
Linfen, Shanxi Province
Shanxi Provincial Institute of Archaeology

铜 铃

Chariot bell

西周（公元前 1046 年－公元前 771 年）
高 10.5 厘米，铣间距 8.6 厘米
山西省临汾市翼城县大河口墓地出土
山西省考古研究所藏

Western Zhou Dynasty (1046 B.C. –771 B.C.)
Height: 10.5 cm, arc spacing: 8.6 cm
Unearthed from Dahekou tomb complex in
Yicheng County, Linfen, Shanxi Province
Shanxi Provincial Institute of Archaeology

铜 铃

Chariot bell

西周（公元前 1046 年－公元前 771 年）
高 8.2 厘米，铣间距 6.2 厘米
山西省临汾市翼城县大河口墓地出土
山西省考古研究所藏

Western Zhou Dynasty (1046 B.C. –771 B.C.)
Height: 8.2 cm, arc spacing: 6.2 cm
Unearthed from Dahekou tomb complex in
Yicheng County, Linfen, Shanxi Province
Shanxi Provincial Institute of Archaeology

　　西周时期，玉器在服制和礼制中都有举足轻重的地位，既有礼玉的性质，又有装饰的功能。随着其结构的复杂化和制度化，逐渐成为权贵身份的象征。《诗经》中的"佩玉将将"，是说走起路来玉器相撞，发出互相碰击的铿锵之声，从而显现出佩戴者的身份；同时在行走时倾听玉声，联想玉德，提醒自己恪守礼制。

大河口墓地青铜器和玉器出土照片

二联璜组玉佩

Jade ornaments

西周（公元前 1046 年 - 公元前 771 年）
长 35.3 厘米，最宽 16 厘米
山西省临汾市翼城县大河口墓地出土
山西省考古研究所藏

Western Zhou Dynasty (1046 B.C. −771 B.C.)
Length: 35.3 cm, largest width: 16 cm
Unearthed from Dahekou tomb complex in
Yicheng County, Linfen, Shanxi Province
Shanxi Provincial Institute of Archaeology

　　1 组 40 件，包括玉堵头 1 件，大
小相同的玉璜 2 件，小蚕蛹 11 件，余
为红玛瑙和绿松石管。

《诗经·小雅·楚茨》

济济跄跄，洁尔牛羊，以往烝尝。

或剥或亨，或肆或将。

祝祭于祊，祀事孔明。

先祖是皇，神保是飨。

孝孙有庆，报以介福，万寿无疆"。

Sacrificial Ceremony:
Important Event
of the State

祭祀

国之大事

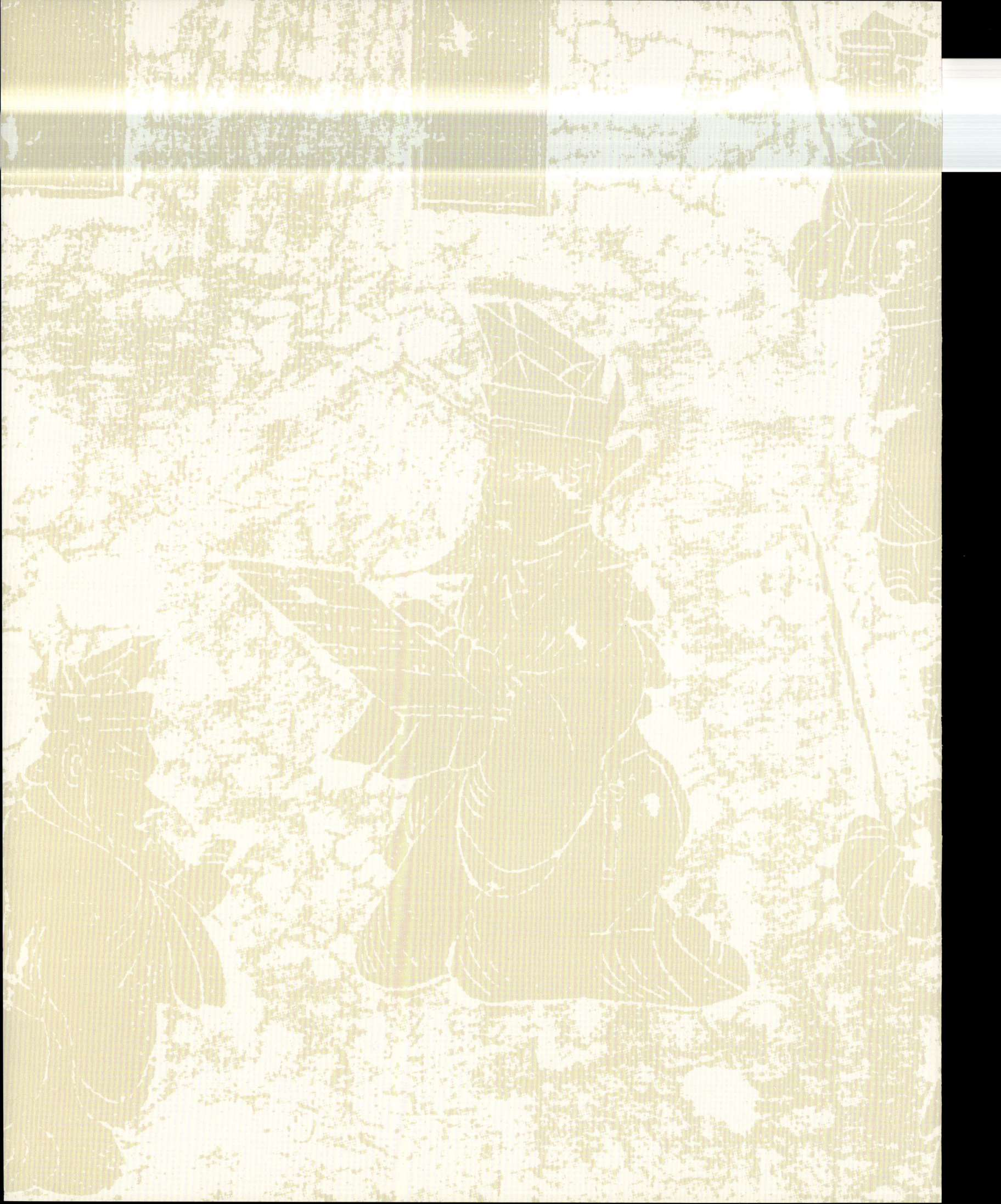

霸伯把报答祖先父母之恩德、感谢天地自然之恩惠作为祭祀的目的，并将之与战争一起列为国家的大事，这不仅成为了霸国人遵守的道德准则，而且上升为国家制度。

The monarch of Ba carried out the sacrificial ceremony for the purpose of showing gratitude to the ancestors, parents, earth, heaven and nature for having attained the blessedness of those benefactors. The ceremony itself was considered as an event as crucial as a military operation. Not only had it been codified to be one of the most basic moral principles but also one of the state institutions that the people of Ba should have to comply with.

礼通常是由礼器的大小、多少、繁简等来表示礼数的高低。

一、礼有以多为贵者，宗庙之数，天子七庙，诸侯五庙，大夫三庙，士一庙。行礼时盛食用的豆，天子二十六，诸公十六，诸侯十二，上大夫八，下大夫六。

二、礼有以高为贵者，如天子之堂九尺，诸侯七尺，大夫五尺，士三尺。

三、礼有以大为贵者，宫室、器皿、丘封等，都以大为贵，棺椁也以厚为贵。

四、礼有以文为贵者，愈尊者，文饰愈复杂。

五、礼有以小为贵者，宗庙之祭，贵者献以爵，贱者献以散；尊者举觯（音质），卑者举角。爵的容量为一升，散为五升，所以前贵后贱。觯的容量为三升，角为四升，所以前尊后卑。

六、礼有以素为贵者，至敬无文，父党无容，大圭不琢，大羹不和。大圭是天子祭祀时插在绅带之间的玉器，或称为珽，不加雕琢。大羹是煮肉汁，不加盐菜，不致五味。

七、礼有以少为贵者，如天子祭天，天神至尊无二，所以天子祭天用"特牲"，即一头牛。诸侯侍奉天子，犹如天子事天，故天子巡视到诸侯境内时，诸侯也以一牛为膳进献之。食礼有劝食，天子仅一食即告饱，诸侯再食，大夫三食，原因是尊者常以德为饱，不以食味为重，诸侯、大夫之德递降，所以食数也随之递增。

金柄形器

Gold ritual ware

西周（公元前 1046 年－公元前 771 年）
长 11 厘米，宽 2.2 厘米，厚 0.3 厘米
山西省临汾市翼城县大河口墓地出土
山西省考古研究所藏

Western Zhou Dynasty (1046 B.C. −771 B.C.)
Length: 11 cm, width: 2.2 cm, thickness: 0.3 cm
Unearthed from Dahekou tomb complex in Yicheng County,
Linfen, Shanxi Province
Shanxi Provincial Institute of Archaeology

据推测应为圭。其黄金含量达到近 84%。

金 璜

Gold *huang* ritual ware

西周（公元前 1046 年－公元前 771 年）
长 5.5 厘米，宽 2.5 厘米
山西省临汾市翼城县大河口墓地出土
山西省考古研究所藏

Western Zhou Dynasty (1046 B.C. −771 B.C.)
Length: 5.5 cm, width: 2.5 cm
Unearthed from Dahekou tomb complex in Yicheng
County, Linfen, Shanxi Province
Shanxi Provincial Institute of Archaeology

"圭璋特，琥璜爵"，圭、璋都是礼器中
的贵重者，在礼仪活动中，可单独作为信物使用;
琥璜的重要性次于圭璋，在天子宴诸侯或诸侯
相宴时，与爵同时进上。这件金璜是在对盗洞
内的回填土过筛时发现的。

圆涡纹罍

Lei wine container with paisley pattern

西周（公元前 1046 年－公元前 771 年）

高 38.6 厘米，口径 16.4 厘米，耳间距 35.5 厘米

山西省临汾市翼城县大河口墓地出土

山西省考古研究所藏

Western Zhou Dynasty (1046 B.C. –771 B.C.)
Height: 38.6 cm, mouth diameter: 16.4 cm, distance between ears: 35.5 cm
Unearthed from Dahekou tomb complex in Yicheng County, Linfen, Shanxi
Province
Shanxi Provincial Institute of Archaeology

　　罍，音雷。大型的盛酒器，又可盛水，在青铜礼器中占有重要的地位。《周礼·春官》载："凡祭祀……用大罍。"罍起源于商代晚期，流行于西周和春秋。

凤鸟纹尊

Zun wine container with phoenix pattern

西周（公元前 1046 年 – 公元前 771 年）

高 25.4 厘米，口径 21.1 厘米，底径 17.1 厘米

山西省临汾市翼城县大河口墓地出土

山西省考古研究所藏

Western Zhou Dynasty (1046 B.C. –771 B.C.)
Height: 25.4 cm, mouth diameter: 21.1 cm, base
diameter: 17.1 cm
Unearthed from Dahekou tomb complex in Yicheng
County, Linfen, Shanxi Province
Shanxi Provincial Institute of Archaeology

盛酒器。器内底壁有铭文四字。

兽面纹方鼎

Square *ding* cauldron with beast-face design

西周（公元前 1046 年 - 公元前 771 年）

高 20 厘米，耳间距 16.6 厘米

山西省临汾市翼城县大河口墓地出土

山西省考古研究所藏

Western Zhou Dynasty (1046 B.C. –771 B.C.)
Height: 20 cm, distance between ears: 16.6 cm
Unearthed from Dahekou tomb complex in Yicheng County,
Linfen, Shanxi Province
Shanxi Provincial Institute of Archaeology

兽面纹方鼎

Square *ding* cauldron with beast-face design

西周（公元前 1046 年 – 公元前 771 年）

高 20.8 厘米，耳间距 16.7 厘米

山西省临汾市翼城县大河口墓地出土

山西省考古研究所藏

Western Zhou Dynasty (1046 B.C. –771 B.C.)
Height: 20.8 cm, distance between ears: 16.7 cm
Unearthed from Dahekou tomb complex in Yicheng County,
Linfen, Shanxi Province
Shanxi Provincial Institute of Archaeology

火纹四目纹鼎

Ding tripod cauldron with fire emblem
and four eyes pattern

西周（公元前 1046 年－公元前 771 年）

高 20.4 厘米，耳间距 17 厘米

山西省临汾市翼城县大河口墓地出土

山西省考古研究所藏

Western Zhou Dynasty (1046 B.C. –771 B.C.)
Height: 20.4 cm, distance between ears: 17 cm
Unearthed from Dahekou tomb complex in Yicheng
County, Linfen, Shanxi Province
Shanxi Provincial Institute of Archaeology

龙纹鼎

Ding tripod cauldron with dragon design

西周（公元前 1046 年 – 公元前 771 年）

高 16.6 厘米，耳间距 16 厘米

山西省临汾市翼城县大河口墓地出土

山西省考古研究所藏

Western Zhou Dynasty (1046 B.C. –771 B.C.)
Height: 16.6 cm, distance between ears: 16 cm
Unearthed from Dahekou tomb complex in Yicheng
County, Linfen, Shanxi Province
Shanxi Provincial Institute of Archaeology

器外底有烟灰痕迹。

龙纹鼎

Ding tripod cauldron with dragon design

西周（公元前 1046 年 – 公元前 771 年）

高 13.3 厘米，耳间距 13 厘米

山西省临汾市翼城县大河口墓地出土

山西省考古研究所藏

Western Zhou Dynasty (1046 B.C. –771 B.C.)
Height: 13.3 cm, distance between ears: 13 cm
Unearthed from Dahekou tomb complex in Yicheng
County, Linfen, Shanxi Province
Shanxi Provincial Institute of Archaeology

凤鸟纹簋

Gui food container with phoenix design

西周（公元前 1046 年 – 公元前 771 年）

高 13.7 厘米，口径 14.3 厘米，耳间距 26 厘米

山西省临汾市翼城县大河口墓地出土

山西省考古研究所藏

Western Zhou Dynasty (1046 B.C. –771 B.C.)
Height: 13.7 cm, distance between ears: 26 cm, mouth diameter: 14.3 cm
Unearthed from Dahekou tomb complex in Yicheng County, Linfen, Shanxi Province
Shanxi Provincial Institute of Archaeology

兽耳龙纹簋

Gui food container with design of beast-shaped ears and dragon pattern

西周（公元前 1046 年－公元前 771 年）

高 16.5 厘米，口径 24 厘米，耳间距 33 厘米

山西省临汾市翼城县大河口墓地出土

山西省考古研究所藏

Western Zhou Dynasty (1046 B.C. –771 B.C.)
Height: 16.5 cm, mouth diameter: 24 cm, distance between ears: 33 cm
Unearthed from Dahekou tomb complex in Yicheng County, Linfen,
Shanxi Province
Shanxi Provincial Institute of Archaeology

器底内壁有铭文"伯口肇乍鼎彝子子孙孙永宝"，乍即作。

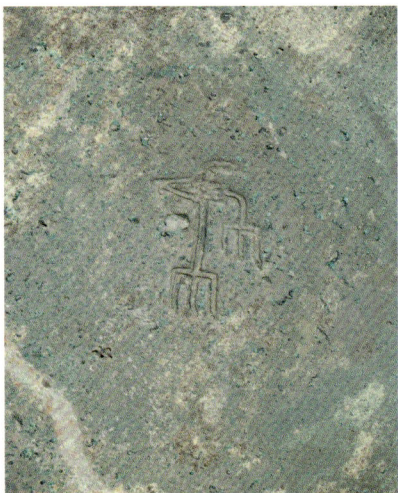

兽耳火纹簋

Gui food container with design of beast-shaped ears and fire emblem

西周（公元前 1046 年 – 公元前 771 年）
高 14 厘米，口径 21 厘米，耳间距 27 厘米
山西省临汾市翼城县大河口墓地出土
山西省考古研究所藏

Western Zhou Dynasty (1046 B.C. –771 B.C.)
Height: 14 cm, mouth diameter: 21 cm, distance between ears: 27 cm
Unearthed from Dahekou tomb complex in Yicheng County, Linfen, Shanxi Province
Shanxi Provincial Institute of Archaeology

斜线纹鬲

Li steamer with diagonal lines pattern

西周（公元前 1046 年 – 公元前 771 年）

高 12.4 厘米，耳间距 13.5 厘米

山西省临汾市翼城县大河口墓地出土

山西省考古研究所藏

Western Zhou Dynasty (1046 B.C. –771 B.C.)
Height: 12.4 cm, distance between ears: 13.5 cm
Unearthed from Dahekou tomb complex in
Yicheng County, Linfen, Shanxi Province
Shanxi Provincial Institute of Archaeology

炊煮器。

斜线纹鬲

Li steamer with diagonal lines pattern

西周（公元前 1046 年 – 公元前 771 年）

高 11.5 厘米，耳间距 15.5 厘米

山西省临汾市翼城县大河口墓地出土

山西省考古研究所藏

Western Zhou Dynasty (1046 B.C. –771 B.C.)
Height: 11.5 cm, distance between ears: 15.5 cm
Unearthed from Dahekou tomb complex in
Yicheng County, Linfen, Shanxi Province
Shanxi Provincial Institute of Archaeology

铜甗

Yan steamer

西周（公元前 1046 年 - 公元前 771 年）
高 46.5 厘米，耳间距 33.5 厘米
山西省临汾市翼城县大河口墓地出土
山西省考古研究所藏

Western Zhou Dynasty (1046 B.C. – 771 B.C.)
Height: 46.5 cm, distance between ears: 33.5 cm
Unearthed from Dahekou tomb complex in
Yicheng County, Linfen, Shanxi Province
Shanxi Provincial Institute of Archaeology

铜提梁卣

You wine container with overtop handle

西周（公元前 1046 年 - 公元前 771 年）
高 28.5 厘米，耳间距 20 厘米，腹径 18 厘米
山西省临汾市翼城县大河口墓地出土
山西省考古研究所藏

Western Zhou Dynasty (1046 B.C. –771 B.C.)
Height: 28.5 cm, distance between ears: 20 cm,
belly diameter: 18 cm
Unearthed from Dahekou tomb complex in
Yicheng County, Linfen, Shanxi Province
Shanxi Provincial Institute of Archaeology

　　卣，盛酒器。流行于商晚期和西周早期，
祭祀时用于盛放名为"秬鬯"（音具畅）的香酒。
秬鬯是以黑黍和郁金香草酿造的酒，用于祭祀
降神。

铜提梁卣

You wine container with overtop handle

西周（公元前 1046 年 - 公元前 771 年）
高 22.8 厘米，耳间距 16 厘米，腹径 18 厘米 ×13 厘米
山西省临汾市翼城县大河口墓地出土
山西省考古研究所藏

Western Zhou Dynasty (1046 B.C. –771 B.C.)
Height: 22.8 cm, distance between ears: 16 cm, belly
diameter: 18 cm × 13 cm
Unearthed from Dahekou tomb complex in Yicheng
County, Linfen, Shanxi Province
Shanxi Provincial Institute of Archaeology

陶爵杯及原始瓷器出土照片

陶爵杯

Ceramic *jue* cups

西周（公元前 1046 年 - 公元前 771 年）
上：高 17 厘米，口径 13.5 厘米，把与口
沿间长 24 厘米，足径 9.8 厘米
下：高 14 厘米，口径 11.7 厘米，把与口
沿间长 18.6 厘米，足径 9.2 厘米
山西省临汾市翼城县大河口墓地出土
山西省考古研究所藏

Western Zhou Dynasty (1046 B.C. –771 B.C.)
Upper one: height: 17 cm, mouth diameter: 13.5
cm, distance between handle and mouth rim: 24
cm, bottom diameter: 9.8 cm
Lower one: height: 14 cm, mouth diameter: 11.7
cm, distance between handle and mouth rim: 18.6
cm, bottom diameter: 9.2 cm
Unearthed from Dahekou tomb complex in
Yicheng County, Linfen, Shanxi Province
Shanxi Provincial Institute of Archaeology

　　盛酒器，均带盖，在腹部伸出一个宽
扁的曲柄，底下有碗底一样的圈足。史书
称瓒、爵或废爵，考古学称单把杯或单柄杯。
陕西曾发现一对西周晚期的"伯公父"爵，
在其曲柄的表面有一篇铭文，自名"爵"。
这两件陶器放置在三件瓷器的前面两侧，
是一种有意识的放置，在举行一系列的祭
奠活动中应具有配合使用的功能。

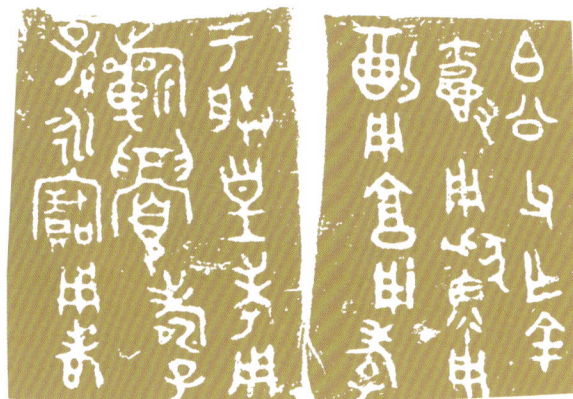

陕西发现的一对西周晚期的"伯公父"爵，铭文为："伯公父作金爵，用献、用酌、用享、用孝于朕皇考。用祈眉寿，子孙永宝用考。"意即伯公父制作金爵，用来祭祀死去的伟大祖先，以求长寿，愿后代子孙万年永宝此物。

原始瓷瓶

Proto-porcelain jar

西周（公元前 1046 年 – 公元前 771 年）
高 13.5 厘米，口径 8.4 厘米，腹径 16.2 厘米，
底径 9.8 厘米
山西省临汾市翼城县大河口墓地出土
山西省考古研究所藏

Western Zhou Dynasty (1046 B.C. –771 B.C.)
Height: 13.5 cm, mouth diameter: 8.4 cm, belly
diameter: 16.2 cm, bottom diameter: 9.8 cm
Unearthed from Dahekou tomb complex in
Yicheng County, Linfen, Shanxi Province
Shanxi Provincial Institute of Archaeology

盛酒器。

原始瓷尊

Proto-porcelain *zun* wine container

西周（公元前 1046 年 – 公元前 771 年）
高 13.9 厘米，口径 12.4 厘米，底径 7.1 厘米
山西省临汾市翼城县大河口墓地出土
山西省考古研究所藏

Western Zhou Dynasty (1046 B.C. –771 B.C.)
Height: 13.9 cm, mouth diameter: 12.4 cm, bottom
diameter: 7.1 cm
Unearthed from Dahekou tomb complex in Yicheng
County, Linfen, Shanxi Province
Shanxi Provincial Institute of Archaeology

Proto-porcelain *dou* food containers

西周（公元前 1046 年 – 公元前 771 年）

高 7.1–9 厘米，最大腹径 10.5–13 厘米，

足径 5.7–8.3 厘米

山西省临汾市翼城县大河口墓地出土

山西省考古研究所藏

Western Zhou Dynasty (1046 B.C. –771 B.C.)
Height: 7.1-9 cm, largest belly diameter: 10.5-13
cm, foot diameter: 5.7 -8.3cm
Unearthed from Dahekou tomb complex in
Yicheng County, Linfen, Shanxi Province
Shanxi Provincial Institute of Archaeology

陶豆（一组）

Pottery *dou* food containers

西周（公元前 1046 年 – 公元前 771 年）

高 11.2–12.3 厘米，口径 13.9–14.5 厘米，

底径 7.2–8.5 厘米

山西省临汾市翼城县大河口墓地出土

山西省考古研究所藏

Western Zhou Dynasty (1046 B.C. –771 B.C.)
Height: 11.2-12.3 cm, mouth diameter: 13.9–14.5
cm, bottom diameter: 7.2-8.5 cm
Unearthed from Dahekou tomb complex in
Yicheng County, Linfen, Shanxi Province
Shanxi Provincial Institute of Archaeology

陶 盘

Pottery *pan* food container

西周（公元前 1046 年 – 公元前 771 年）
高 12.5 厘米，口径 18 厘米，底座径 13.3 厘米
山西省临汾市翼城县大河口墓地出土
山西省考古研究所藏

Western Zhou Dynasty (1046 B.C. –771 B.C.)
Height: 12.5 cm, mouth diameter: 18 cm, base diameter: 13.3 cm
Unearthed from Dahekou tomb complex in Yicheng County, Linfen, Shanxi Province
Shanxi Provincial Institute of Archaeology

盛食器。

磨光陶壶

Pottery pot with polished surface

西周（公元前 1046 年 – 公元前 771 年）

高 29 厘米，腹径 12.3 厘米，足径 12.5 厘米

山西省临汾市翼城县大河口墓地出土

山西省考古研究所藏

Western Zhou Dynasty (1046 B.C. –771 B.C.)
Height: 29 cm, belly diameter: 12.3 cm, base diameter:
12.5 cm
Unearthed from Dahekou tomb complex in Yicheng
County, Linfen, Shanxi Province
Shanxi Provincial Institute of Archaeology

　　盛酒器。

祭祀是霸国礼仪活动中的头等大事，所以遇到重大典礼节日，或旱涝饥荒均要用玉来祭奠，以祈求神灵保佑风调雨顺、国家安定。《诗经》中有"靡神不举"，是说各路神灵都要被供祭。对礼玉精益求精的雕琢和详细的分类，表现出其在祭祀活动中的重要地位和君王对祭礼的恭敬，以及社会对礼制的重视和遵从。

大河口墓地中有一座大墓随葬的青铜器有鼎3件、簋1件、甗1件、盘1件、盆1件、钟1件等，陶器有鬲1件、罐1件、三足瓮1件、未烧结的陶器7件，还有玉石串饰7件、项饰2件、玉玦8件、握玉2件、柄形饰2组。另外还发现了蚌器和贝等。在出土铜甗内壁发现有铸铭"唯正月初吉霸□伯作宝甗其永用"。

墓葬随葬品照片

玉石串饰出土照片

三联璜组玉佩

Jade ornaments

西周（公元前 1046 年－公元前 771 年）

长 37 厘米

山西省临汾市翼城县大河口墓地出土

山西省考古研究所藏

Western Zhou Dynasty (1046 B.C. –771 B.C.)
Length: 37 cm
Unearthed from Dahekou tomb complex in
Yicheng County, Linfen, Shanxi Province
Shanxi Provincial Institute of Archaeology

　　1 组 82 件，由 3 件玉璜、1 件玉堵
头和绿松石及玛瑙管组成。

玉项饰

Jade neck ornaments

西周（公元前 1046 年 - 公元前 771 年）

长 24 厘米，宽 17.5 厘米

山西省临汾市翼城县大河口墓地出土

山西省考古研究所藏

Western Zhou Dynasty (1046 B.C. –771 B.C.)
Length: 24 cm, width: 17.5 cm
Unearthed from Dahekou tomb complex in
Yicheng County, Linfen, Shanxi Province
Shanxi Provincial Institute of Archaeology

　　1 组 8 件，包括 6 件马蹄形器和 2 件素
面璜。出土时位于死者颈部。

玉 斧

Jade axe head

西周（公元前 1046 年 – 公元前 771 年）
长 10.3 厘米，宽 8.3 厘米，厚 1.3 厘米
山西省临汾市翼城县大河口墓地出土
山西省考古研究所藏

Western Zhou Dynasty (1046 B.C. –771 B.C.)
Length: 10.3 cm, width: 8.3 cm, thickness: 1.3 cm
Unearthed from Dahekou tomb complex in Yicheng
County, Linfen, Shanxi Province
Shanxi Provincial Institute of Archaeology

玉璜形器

Semi-annular jade pendant

西周（公元前 1046 年 – 公元前 771 年）
长 8.8 厘米，宽 3.7 厘米，厚 0.5 厘米
山西省临汾市翼城县大河口墓地出土
山西省考古研究所藏

Western Zhou Dynasty (1046 B.C. –771 B.C.)
Length: 8.8 cm, width: 3.7 cm, thickness: 0.5 cm
Unearthed from Dahekou tomb complex in
Yicheng County, Linfen, Shanxi Province
Shanxi Provincial Institute of Archaeology

玉柄形器

Jade ritual ware

西周（公元前 1046 年－公元前 771 年）
高 14.3 厘米，宽 2.5 厘米，厚 1.6 厘米
山西省临汾市翼城县大河口墓地出土
山西省考古研究所藏

Western Zhou Dynasty (1046 B.C. –771 B.C.)
Length: 14.3 cm, width: 2.5 cm, thickness: 1.6 cm
Unearthed from Dahekou tomb complex in
Yicheng County, Linfen, Shanxi Province
Shanxi Provincial Institute of Archaeology

玉柄形器

Jade ritual ware

西周（公元前 1046 年 - 公元前 771 年）

长 10.2 厘米，最大直径 0.9 厘米

山西省临汾市翼城县大河口墓地出土

山西省考古研究所藏

Western Zhou Dynasty (1046 B.C. −771 B.C.)
Length: 10.2 cm, largest diameter: 0.9 cm
Unearthed from Dahekou tomb complex in
Yicheng County, Linfen, Shanxi Province
Shanxi Provincial Institute of Archaeology

玉柄形器

Jade ritual ware

西周（公元前 1046 年 - 公元前 771 年）

长 10.5 厘米，最大直径 1.1 厘米

山西省临汾市翼城县大河口墓地出土

山西省考古研究所藏

Western Zhou Dynasty (1046 B.C. −771 B.C.)
Length: 10.5 cm, Largest diameter: 1.1 cm
Unearthed from Dahekou tomb complex in
Yicheng County, Linfen, Shanxi Province
Shanxi Provincial Institute of Archaeology

Western Zhou Dynasty (1046 B.C. −771 B.C.)
Length: 23.5 cm, width: 7.6 cm, thickness: 0.4 cm
Unearthed from Dahekou tomb complex in
Yicheng County, Linfen, Shanxi Province
Shanxi Provincial Institute of Archaeology

玉 戚

Jade *qi* battle-axe head

西周（公元前 1046 年 – 公元前 771 年）
长 15.2 厘米，宽 7.3 厘米，厚 0.3 厘米
山西省临汾市翼城县大河口墓地出土
山西省考古研究所藏

Western Zhou Dynasty (1046 B.C. –771 B.C.)
Length: 15.2 cm, width: 7.3 cm, thickness: 0.3 cm
Unearthed from Dahekou tomb complex in
Yicheng County, Linfen, Shanxi Province
Shanxi Provincial Institute of Archaeology

玉 戚

Jade *qi* battle-axe head

西周（公元前 1046 年 – 公元前 771 年）
长 7.9 厘米，宽 7.3 厘米，厚 0.2 厘米
山西省临汾市翼城县大河口墓地出土
山西省考古研究所藏

Western Zhou Dynasty (1046 B.C. 771 B.C.)
Length: 7.9 cm, width: 7.3 cm, thickness: 0.2 cm
Unearthed from Dahekou tomb complex in
Yicheng County, Linfen, Shanxi Province
Shanxi Provincial Institute of Archaeology

《诗经·秦风·黄鸟》

交交黄鸟，止于棘。
谁从穆公？子车奄息。
维此奄息，百夫之特。
临其穴，惴惴其栗。
彼苍者天，歼我良人。
如可赎兮，人百其身！

**Funeral Ceremony:
Honoring the Dead
as the Living**

事死如生

丧葬

霸国的丧葬观念正如《荀子·礼论》中所说："丧礼者，以生者饰死者也，大象其生，以送其死，事死如生，事亡如存"，文中将"事死如生"作为丧葬的原则，对待死者如他活着之时，"事"为侍奉、供奉之意，表达了霸国人对生命的敬畏。

We may understand the principle idea pertaining to funeral ceremony from the *Discussion of Ritual Propriety* in the *Book of Xunzi* is that to honor the man during his life and after his death, and to honor the dead as he is still living. This is a concept developed and believed by the Ba people who expressed their reverence for life.

大河口墓地棺盖上
及棺椁间的器物

在椁盖板上发现
了大量的海贝和
青铜车马器

　　山西省翼城县大河口墓葬群中的一座大墓，墓口长 5 米、宽 3.4 米、深 10 米，口小底大，墓主头向西，有腰坑。墓室内发现大量青铜器、玉石器、蚌器、贝等，青铜器种类有食器、酒器、水器、乐器、兵器、工具、车马器等，另外还有金柄形器 1 件，锡器 6 件，陶鬲 1 件。在青铜盂、簋、豆等器内发现铭文，其中铜豆盖内铭"霸伯作大宝尊彝其孙孙子子万年永用"。

　　海贝是中国最早的古代货币，由海贝串成的饰品象征着财富与地位。这座大墓的棺盖板上发现了约 2 万枚海贝，这在西周墓葬中还是第一次发现。海贝的排列似乎有一定的规律，这么多的海贝出现在棺盖板上究竟作何用途呢？现在还难以确定。

赗赙制度：赗赙，音奉富。两周时期贵族死后，由王、诸侯赠送随葬之物，以帮助埋葬死者时使用。《春秋公羊传》有记载："车马曰赗，货财曰赙，衣被曰禭。"霸国墓地中出现倗国的青铜器，就是这一制度的体现。东周时期，礼崩乐坏，宗族制度开始瓦解，各诸侯为了和战，会利用赗赙往来进行外交活动。随着诸侯越来越少，赗赙制度也逐渐退出历史舞台，但却没有消亡，而是转化为民间习俗，至今仍有流传。

凤鸟纹盆

Basins with phoenix pattern

西周（公元前 1046 年 – 公元前 771 年）

高 13 厘米，口径 20 厘米

山西省临汾市翼城县大河口墓地出土

山西省考古研究所藏

Western Zhou Dynasty (1046 B.C. –771 B.C.)
Height: 13 cm, mouth diameter: 20 cm
Unearthed from Dahekou tomb complex in Yicheng County, Linfen, Shanxi Province
Shanxi Provincial Institute of Archaeology

　　这两件青铜盆形状、纹饰和铭文完全相同，应是一对器物。内底面各有一篇相同的铭文："倗伯肇作旅盆，其万年永用"，可知这两件器物是倗伯做的。倗国国君制作的器物出现在霸国墓地，可能是霸伯死后倗伯赠送给霸伯的"赗赙"之物，也就是助丧的礼器，这说明倗国和霸国这两个小国家之间存在着友好的交往关系。

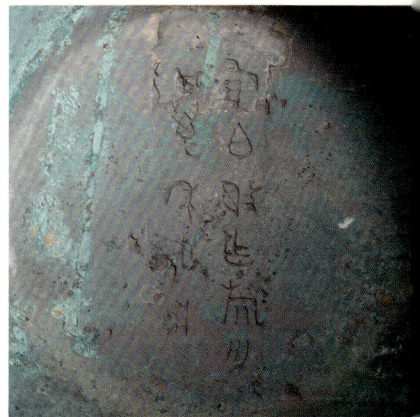

凤鸟纹盆出土照片

铜 爵

Jue wine vessel

西周（公元前 1046 年 – 公元前 771 年）
高 13 厘米，流尾长 10.3 厘米
山西省临汾市翼城县大河口墓地出土
山西省考古研究所藏

Western Zhou Dynasty (1046 B.C. –771 B.C.)
Height: 13 cm, length of spout: 10.3 cm
Unearthed from Dahekou tomb complex in Yicheng
County, Linfen, Shanxi Province
Shanxi Provincial Institute of Archaeology

铜 鼎

Ding tripod cauldron

西周（公元前 1046 年 – 公元前 771 年）
高 11.5 厘米，耳间距 14 厘米
山西省临汾市翼城县大河口墓地出土
山西省考古研究所藏

Western Zhou Dynasty (1046 B.C. –771 B.C.)
Height: 11.5 cm, distance between ears: 14 cm
Unearthed from Dahekou tomb complex in Yicheng
County, Linfen, Shanxi Province
Shanxi Provincial Institute of Archaeology

铜 鼎

Ding tripod cauldron

西周（公元前 1046 年 – 公元前 771 年）
高 13.6 厘米，耳间距 13.7 厘米
山西省临汾市翼城县大河口墓地出土
山西省考古研究所藏

Western Zhou Dynasty (1046 B.C. –771 B.C.)
Height: 13.6 cm, distance between ears: 13.7 cm
Unearthed from Dahekou tomb complex in Yicheng
County, Linfen, Shanxi Province
Shanxi Provincial Institute of Archaeology

四足长方形甗

Rectangular shaped *yan* steamer with four legs

西周（公元前 1046 年 – 公元前 771 年）
高 18 厘米，耳间距 13.5 厘米
山西省临汾市翼城县大河口墓地出土
山西省考古研究所藏

Western Zhou Dynasty (1046 B.C. –771 B.C.)
Height: 18 cm, distance between ears: 13.5 cm
Unearthed from Dahekou tomb complex in Yicheng
County, Linfen, Shanxi Province
Shanxi Provincial Institute of Archaeology

丧葬
事死如生

铜簋及铜盉出土位置图

铜 簋

Gui food container

西周（公元前 1046 年 - 公元前 771 年）
高 11.5 厘米，腹径 12 厘米
山西省临汾市翼城县大河口墓地出土
山西省考古研究所藏

Western Zhou Dynasty (1046 B.C. –771 B.C.)
Height: 11.5 cm, belly diameter: 12 cm
Unearthed from Dahekou tomb complex in Yicheng
County, Linfen, Shanxi Province
Shanxi Provincial Institute of Archaeology

铜 盉

He wine container

西周（公元前 1046 年 - 公元前 771 年）
高 12 厘米，宽 12 厘米，厚 2.5 厘米
山西省临汾市翼城县大河口墓地出土
山西省考古研究所藏

Western Zhou Dynasty (1046 B.C. –771 B.C.)
Height: 12 cm, width: 12 cm, thickness: 2.5 cm
Unearthed from Dahekou tomb complex in Yicheng
County, Linfen, Shanxi Province
Shanxi Provincial Institute of Archaeology

冥器，不具有实用性。器盖与器身铸为一体。

铜盉

He wine container

西周（公元前 1046 年 – 公元前 771 年）
高 13.5 厘米，宽 14.5 厘米，厚 3.5 厘米
山西省临汾市翼城县大河口墓地出土
山西省考古研究所藏

Western Zhou Dynasty (1046 B.C. –771 B.C.)
Height: 13.5 cm, width: 14.5 cm, thickness: 3.5 cm
Unearthed from Dahekou tomb complex in Yicheng
County, Linfen, Shanxi Province
Shanxi Provincial Institute of Archaeology

铜簋

Gui food container

西周（公元前 1046 年 – 公元前 771 年）
高 7.2 厘米，口径 11.5 厘米，圈足径 9.5 厘米
山西省临汾市翼城县大河口墓地出土
山西省考古研究所藏

Western Zhou Dynasty (1046 B.C. –771 B.C.)
Height: 7.2 cm, mouth diameter: 11.5 cm, foot
diameter: 9.5 cm
Unearthed from Dahekou tomb complex in Yicheng
County, Linfen, Shanxi Province
Shanxi Provincial Institute of Archaeology

铜 鼎

Ding tripod cauldron

西周（公元前 1046 年 - 公元前 771 年）
高 11 厘米，耳间距 11.5 厘米
山西省临汾市翼城县大河口墓地出土
山西省考古研究所藏

Western Zhou Dynasty (1046 B.C. –771 B.C.)
Height: 11 cm, distance between ears: 11.5 cm
Unearthed from Dahekou tomb complex in Yicheng
County, Linfen, Shanxi Province
Shanxi Provincial Institute of Archaeology

带盖簋

Gui food container with a lid

西周（公元前 1046 年 - 公元前 771 年）
高 13 厘米，耳间距 20 厘米
山西省临汾市翼城县大河口墓地出土
山西省考古研究所藏

Western Zhou Dynasty (1046 B.C. –771 B.C.)
Height: 13 cm, distance between ears: 20 cm
Unearthed from Dahekou tomb complex in Yicheng
County, Linfen, Shanxi Province
Shanxi Provincial Institute of Archaeology

铜 盉

He wine container

西周（公元前 1046 年 – 公元前 771 年）

高 9 厘米，口径 7.7 厘米

山西省临汾市翼城县大河口墓地出土

山西省考古研究所藏

Western Zhou Dynasty (1046 B.C. –771 B.C.)
Height: 9 cm, mouth diameter: 7.7 cm
Unearthed from Dahekou tomb complex in Yicheng
County, Linfen, Shanxi Province
Shanxi Provincial Institute of Archaeology

铜 盘

Pan basin

西周（公元前 1046 年 – 公元前 771 年）

高 5.9 厘米，耳间距 17.5 厘米，圈足径 11.2 厘米

山西省临汾市翼城县大河口墓地出土

山西省考古研究所藏

Western Zhou Dynasty (1046 B.C. –771 B.C.)
Height: 5.9 cm, distance between ears: 17.5 cm, foot diameter: 11.2 cm
Unearthed from Dahekou tomb complex in Yicheng County, Linfen, Shanxi Province
Shanxi Provincial Institute of Archaeology

丧葬
事死如生

铜 盉

He wine container

西周（公元前 1046 年 – 公元前 771 年）

高 12 厘米，流鋬间距 13 厘米

山西省临汾市翼城县大河口墓地出土

山西省考古研究所藏

Western Zhou Dynasty (1046 B.C. –771 B.C.)
Height: 12 cm, distance between spout and handle:
13 cm
Unearthed from Dahekou tomb complex in Yicheng
County, Linfen, Shanxi Province
Shanxi Provincial Institute of Archaeology

铜 盘

Pan basin

西周（公元前 1046 年 – 公元前 771 年）

高 6.5 厘米，耳间距 21.5 厘米，底径 13.5 厘米

山西省临汾市翼城县大河口墓地出土

山西省考古研究所藏

Western Zhou Dynasty (1046 B.C. –771 B.C.)
Height: 6.5 cm, distance between ears: 21.5 cm, foot
diameter: 13.5 cm
Unearthed from Dahekou tomb complex in Yicheng
County, Linfen, Shanxi Province
Shanxi Provincial Institute of Archaeology

铜 盉

He wine container

西周（公元前 1046 年 – 公元前 771 年）

高 13.2 厘米，流鋬间距 14.5 厘米

山西省临汾市翼城县大河口墓地出土

山西省考古研究所藏

Western Zhou Dynasty (1046 B.C. –771 B.C.)
Height: 13.2 cm, distance between spout and handle:
14.5 cm
Unearthed from Dahekou tomb complex in Yicheng
County, Linfen, Shanxi Province
Shanxi Provincial Institute of Archaeology

陶三足罐

Pottery tripod jar

西周（公元前 1046 年 – 公元前 771 年）

高 20.5 厘米，口径 9.6 厘米，腹径 23 厘米

山西省临汾市翼城县大河口墓地出土

山西省考古研究所藏

Western Zhou Dynasty (1046 B.C. –771 B.C.)
Height: 20.5 cm, mouth diameter: 9.6 cm, belly
diameter: 23 cm
Unearthed from Dahekou tomb complex in Yicheng
County, Linfen, Shanxi Province
Shanxi Provincial Institute of Archaeology

陶罐

Pottery jar

西周（公元前 1046 年 – 公元前 771 年）

高 17 厘米，口径 10.5 厘米，腹径 15.5 厘米

山西省临汾市翼城县大河口墓地出土

山西省考古研究所藏

Western Zhou Dynasty (1046 B.C. −771 B.C.)
Height: 17 cm, mouth diameter: 10.5 cm, belly diameter: 15.5 cm
Unearthed from Dahekou tomb complex in Yicheng County, Linfen, Shanxi Province
Shanxi Provincial Institute of Archaeology

陶 簋

Pottery *gui* food container

西周（公元前 1046 年 – 公元前 771 年）

高 24 厘米，口径 13 厘米，腹径 21 厘米

山西省临汾市翼城县大河口墓地出土

山西省考古研究所藏

Western Zhou Dynasty (1046 B.C. −771 B.C.)
Height: 24 cm, mouth diameter: 13 cm, belly diameter: 21 cm
Unearthed from Dahekou tomb complex in Yicheng County, Linfen, Shanxi Province
Shanxi Provincial Institute of Archaeology

陶 簋

Pottery *gui* food container

西周（公元前 1046 年 - 公元前 771 年）

高 15.7 厘米，口径 21.3 厘米，足径 12.5 厘米

山西省临汾市翼城县大河口墓地出土

山西省考古研究所藏

Western Zhou Dynasty (1046 B.C. – 771 B.C.)
Height: 15.7 cm, mouth diameter: 21.3 cm, foot diameter: 12.5 cm
Unearthed from Dahekou tomb complex in Yicheng County, Linfen, Shanxi Province
Shanxi Provincial Institute of Archaeology

　　盛食器。

陶 鬲

Pottery *li* steamer

西周（公元前 1046 年 - 公元前 771 年）

高 13 厘米，口径 13.2 厘米

山西省临汾市翼城县大河口墓地出土

山西省考古研究所藏

Western Zhou Dynasty (1046 B.C. –771 B.C.)
Height: 13 cm, mouth diameter: 13.2 cm
Unearthed from Dahekou tomb complex in Yicheng County, Linfen, Shanxi Province
Shanxi Provincial Institute of Archaeology

　　炊煮器。

陶 豆

Pottery *dou* food container

西周（公元前 1046 年 – 公元前 771 年）

高 9.2 厘米，口径 15.5 厘米，底径 12.2 厘米

山西省临汾市翼城县大河口墓地出土

山西省考古研究所藏

Western Zhou Dynasty (1046 B.C. –771 B.C.)
Height: 9.2 cm, mouth diameter: 15.5 cm, base
diameter: 12.2 cm
Unearthed from Dahekou tomb complex in Yicheng
County, Linfen, Shanxi Province
Shanxi Provincial Institute of Archaeology

盛食器。

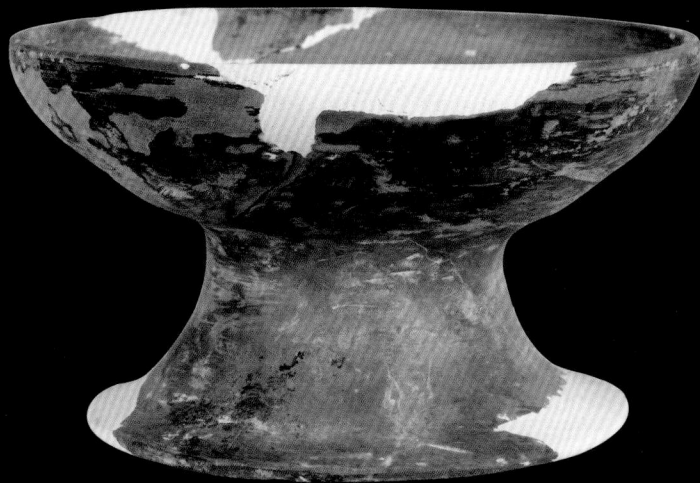

陶 盆

Pottery basin

西周（公元前 1046 年 – 公元前 771 年）

高 16.8 厘米，口径 27 厘米

山西省临汾市翼城县大河口墓地出土

山西省考古研究所藏

Western Zhou Dynasty (1046 B.C. –771 B.C.)
Height: 16.8 cm, mouth diameter: 27 cm
Unearthed from Dahekou tomb complex in Yicheng
County, Linfen, Shanxi Province
Shanxi Provincial Institute of Archaeology

盛器。

大口尊

Zun wine container with large mouth

西周（公元前 1046 年 – 公元前 771 年）

高 26.5 厘米，口径 27.5 厘米

山西省临汾市翼城县大河口墓地出土

山西省考古研究所藏

Western Zhou Dynasty (1046 B.C. –771 B.C.)
Height: 26.5 cm, mouth diameter: 27.5 cm
Unearthed from Dahekou tomb complex in Yicheng
County, Linfen, Shanxi Province
Shanxi Provincial Institute of Archaeology

盛器。

带盖陶壶

Pottery pot with a lid

西周（公元前 1046 年 – 公元前 771 年）

高 26.5 厘米，口径 8 厘米，腹径 15.5 厘米

山西省临汾市翼城县大河口墓地出土

山西省考古研究所藏

Western Zhou Dynasty (1046 B.C. –771 B.C.)
Height: 26.5 cm, mouth diameter: 8 cm, belly
diameter: 15.5 cm
Unearthed from Dahekou tomb complex in Yicheng
County, Linfen, Shanxi Province
Shanxi Provincial Institute of Archaeology

盛器。出土时位于棺外西北部二层台上。

丧葬
事死如生

131

铜 凿

Bronze chisel

西周（公元前 1046 年 – 公元前 771 年）
长 10.5 厘米，宽 2.1 厘米
山西省临汾市翼城县大河口墓地出土
山西省考古研究所藏

Western Zhou Dynasty (1046 B.C. –771 B.C.)
Length: 10.5 cm, width: 2.1 cm
Unearthed from Dahekou tomb complex in
Yicheng County, Linfen, Shanxi Province
Shanxi Provincial Institute of Archaeology

凿木工具。方銎内遗残木柄。

玉 环

Jade disk with a larger orifice

西周（公元前 1046 年 – 公元前 771 年）
直径 13.5 厘米，厚 0.3 厘米，肉宽 3.1 厘米
山西省临汾市翼城县大河口墓地出土
山西省考古研究所藏

Western Zhou Dynasty (1046 B.C. –771 B.C.)
Diameter: 13.5 cm, thickness: 0.3 cm, wall thickness:
3.1 cm
Unearthed from Dahekou tomb complex in Yicheng
County, Linfen, Shanxi Province
Shanxi Provincial Institute of Archaeology

玉覆面

Jade face ornaments

西周（公元前 1046 年 – 公元前 771 年）

长 19.5 厘米，最宽 23 厘米

山西省临汾市翼城县大河口墓地出土

山西省考古研究所藏

Western Zhou Dynasty (1046 B.C. –771 B.C.)
Length: 19.5 cm, largest width: 23 cm
Unearthed from Dahekou tomb complex in
Yicheng County, Linfen, Shanxi Province
Shanxi Provincial Institute of Archaeology

　　1 组 8 件，由 1 件璜、1 件虎、1 件龙、4 件蚕蛹和 1 件鱼组成。根据《仪礼·士丧礼》的记载，在先秦的丧葬仪式中，要用布、帛制成的"幎目"（也称为"覆面"）和"掩"（也叫"裹首"）来包裹死者的头脸部位。古人认为玉可以防止灵魂出壳，保证尸体不腐烂，西周时期这种特殊的丧葬用玉——玉覆面便出现了。它用各种玉料对应人的五官及面部特征制成饰片，缀饰于纺织品上，用于殓葬时覆盖在死者面部，这种奢华的丧葬品也仅出现于贵族墓葬中。

　　玉覆面在西周时已相当流行，其配置格式并非一成不变，常因墓主身份地位或家境情况的不同而有繁有简。由于眼睛在人的五官中最为重要，所以不论怎么简化，眼玉是不能去掉的。

《礼记·檀弓下》："奠以素器，以生者有哀素之心也。唯祭祀之礼，主人自尽焉尔，岂知神之所飨，亦以主人有齐敬之心也。"大意为：供奉死者的酒食，用质朴的器皿盛放，因为生者有悲哀灰冷的心绪。只有埋葬以后的种种祭礼，主人才用有纹饰的器皿，以尽自己敬爱之心。

霸伯墓葬的 11 个壁龛中有漆木俎、角状杯、双耳杯、单把杯、碗、牺尊、豆、罍、壶、龙凤屏风等漆木器，大多为祭祀用礼器；在二层台上出土的漆木器有人俑、盾牌、尊等；棺椁间出土的漆木器有兵器柄、木扳指、木杖等，共计发现漆木器 60 多件。

漆木俎出土照片

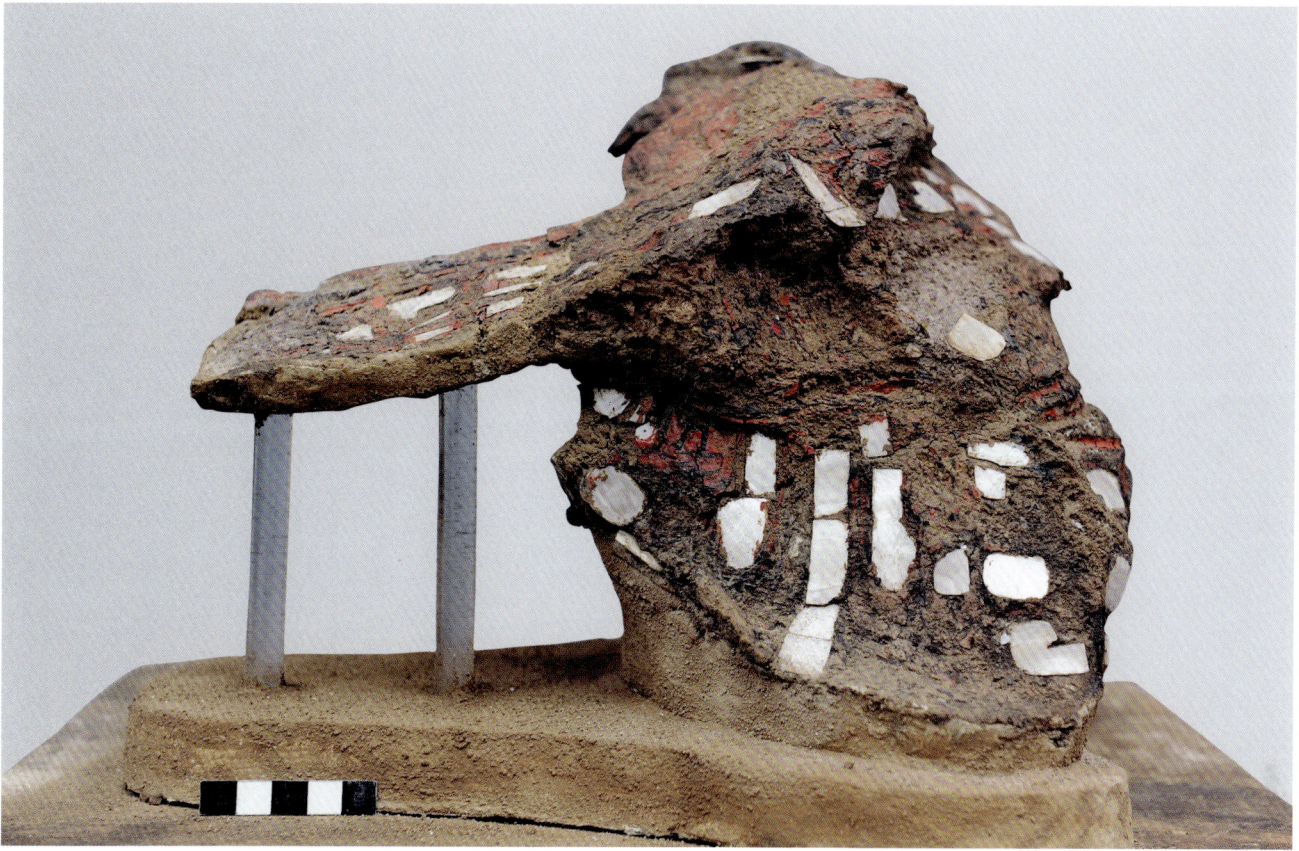

漆木方彝

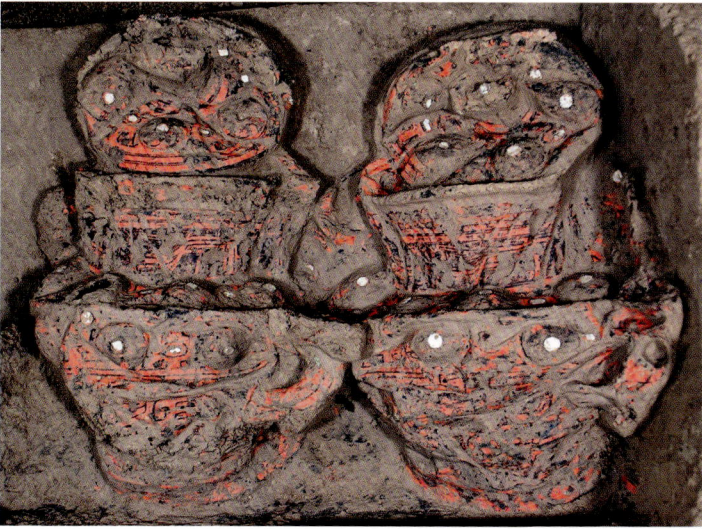

漆木罍及复原图

漆木壶

丧葬
事死如生

135

漆木双耳杯

漆木碗

漆木屏风

漆木人俑

Lacquer Figurines

西周（公元前 1046 年 – 公元前 771 年）
长 205 厘米，宽 73 厘米，高 120 厘米
山西省临汾市翼城县大河口墓地出土
山西省考古研究所藏

Western Zhou Dynasty (1046 B.C. –771 B.C.)
Length: 205 cm, width: 73 cm, height: 120 cm,
Unearthed from Dahekou tomb complex in Yicheng County,
Linfen, Shanxi Province
Shanxi Provincial Institute of Archaeology

　　殷商时期普遍流行活人殉葬，霸国墓葬使用俑，是陪葬制度的重大变革。
这两个漆木人俑，高约 1 米，双足站立于漆木龟上，双手作持物状。据推测其
与礼制和宗教有着微妙的关系。在西周考古史上，墓内随葬漆木人俑这是第一
次发现。此前，发现最早的漆木人俑出自陕西韩城梁带村春秋早期墓葬内。

陕西韩城梁带村发现的彩绘木俑

葬
事死如生

漆木牺尊

Lacquer wine vessel in the shape of a beast

西周（公元前 1046 年 – 公元前 771 年）

长 70 厘米，高 50 厘米，最宽处 25.5 厘米

山西省临汾市翼城县大河口墓地出土

山西省考古研究所藏

Western Zhou Dynasty (1046 B.C. –771 B.C.)
Length: 70 cm, height: 50 cm, largest width: 25.5 cm
Unearthed from Dahekou tomb complex in Yicheng County,
Linfen, Shanxi Province
Shanxi Provincial Institute of Archaeology

盛酒器。

漆木觚

Lacquer wine beakers

西周（公元前 1046 年 – 公元前 771 年）

高约 40 厘米

山西省临汾市翼城县大河口墓地出土

山西省考古研究所藏

Western Zhou Dynasty (1046 B.C. –771 B.C.)
Height: c.a. 40 cm
Unearthed from Dahekou tomb complex in Yicheng
County, Linfen, Shanxi Province
Shanxi Provincial Institute of Archaeology

　　酒器。1 组 2 件，同出于一件青铜尊内。漆木觚口腹内有一个木腔。大河口墓地还出土了一件青铜觚，其内部也有木质内腔，与漆木觚内的木腔相似。

漆木权杖

Lacquer sceptre

西周（公元前 1046 年 – 公元前 771 年）

长 108 厘米，杖首长 32 厘米，宽 8 厘米

山西省临汾市翼城县大河口墓地出土

山西省考古研究所藏

Western Zhou Dynasty (1046 B.C. –771 B.C.)
Length: 108 cm, length of head: 32 cm, width: 8 cm
Unearthed from Dahekou tomb complex in Yicheng County, Linfen,
Shanxi Province
Shanxi Provincial Institute of Archaeology

漆木禁

Lacquer long table

西周（公元前 1046 年 – 公元前 771 年）

禁面长 44 厘米，宽 17 厘米，高 19 厘米

山西省临汾市翼城县大河口墓地出土

山西省考古研究所藏

Western Zhou Dynasty (1046 B.C. –771 B.C.)
Length: 44 cm, width: 17 cm, height: 19 cm
Unearthed from Dahekou tomb complex in Yicheng County,
Linfen, Shanxi Province
Shanxi Provincial Institute of Archaeology

盛放酒器的案子。

漆木豆

Lacquer *dou* food container

西周（公元前 1046 年 – 公元前 771 年）

口径 17.5 厘米，高 18.5 厘米，座径 16 厘米，柄高 8 厘米

复原品：口径 19 厘米，底径 16 厘米，高 19 厘米

山西省临汾市翼城县大河口墓地出土

山西省考古研究所藏

Western Zhou Dynasty (1046 B.C. –771 B.C.)
Mouth diameter: 17.5 cm, height: 18.5 cm, base diameter:
16 cm, height of handle: 8 cm
Unearthed from Dahekou tomb complex in Yicheng County,
Linfen, Shanxi Province
Shanxi Provincial Institute of Archaeology

盛放腌制的菜、肉酱等调味品的器物。

（复原品）

（复原品）

漆木单把杯

Lacquer wine vessel

西周（公元前 1046 年 – 公元前 771 年）
高约 7.3 厘米，口径约 10 厘米，
底径 5.5 厘米
山西省临汾市翼城县大河口墓地出土
山西省考古研究所藏

Western Zhou Dynasty (1046 B.C. –771 B.C.)
Height: 7.3 cm, mouth diameter ca.: 10 cm, base
diameter: 5.5 cm
Unearthed from Dahekou tomb complex in
Yicheng County, Linfen, Shanxi Province
Shanxi Provincial Institute of Archaeology

酒器。

漆木角状杯

Lacquer wine vessel in the shape of a
horn

西周（公元前 1046 年 – 公元前 771 年）
杯高 8 厘米，口径 9 厘米，柄长 13.7 厘米
（象牙件长 5.3 厘米）
山西省临汾市翼城县大河口墓地出土
山西省考古研究所藏

Western Zhou Dynasty (1046 B.C. –771 B.C.)
Height: 8 cm, mouth diameter: 9 cm, length of
handle: 13.7 cm (ivory: 5.3 cm)
Unearthed from Dahekou tomb complex in
Yicheng County, Linfen, Shanxi Province
Shanxi Provincial Institute of Archaeology

酒器。

丧葬
事死如生

143

在墓葬中随葬大量的兵器，象征着霸伯征伐的权力，也与墓主生前的职业经历密不可分。霸国的士兵如同《诗经》里颂扬的"赳赳武夫"一般器宇轩昂，兵器像军功章一样能充分表达他们平生的功绩。另一方面，霸国时虽崇尚武力，却也将弓箭当做竞技活动的工具，在射击比赛中提倡重人道，讲礼仪，守信让，引导社会趋于和谐。这或许是随葬兵器的另一含义吧。

棺盖板上带竹木箭杆的青铜镞

丧葬
事死如生

二层台上的兵器

145

铜短剑

Sword

西周（公元前1046年－公元前771年）

长28.2厘米，最宽4.5厘米

山西省临汾市翼城县大河口墓地出土

山西省考古研究所藏

Western Zhou Dynasty (1046 B.C. –771 B.C.)
Length: 28.2 cm, largest width: 4.5 cm
Unearthed from Dahekou tomb complex in Yicheng
County, Linfen, Shanxi Province
Shanxi Provincial Institute of Archaeology

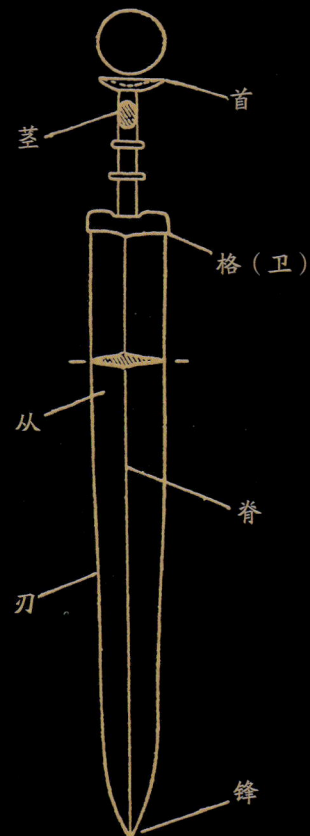

首

茎

格（卫）

从

脊

刃

锋

铜 矛

Spear head

西周（公元前1046年－公元前771年）

长20.3厘米

山西省临汾市翼城县大河口墓地出土

山西省考古研究所藏

Western Zhou Dynasty (1046 B.C. –771 B.C.)
Length: 20.3 cm
Unearthed from Dahekou tomb complex in Yicheng
County, Linfen, Shanxi Province
Shanxi Provincial Institute of Archaeology

含有浓郁的北方青铜文化的气息，弥足珍贵。

铜 矛

Spear head

西周（公元前 1046 年－公元前 771 年）

长 24 厘米，最宽 4.3 厘米

山西省临汾市翼城县大河口墓地出土

山西省考古研究所藏

Western Zhou Dynasty (1046 B.C. –771 B.C.)
Length: 24 cm, largest width: 4.3 cm
Unearthed from Dahekou tomb complex in Yicheng
County, Linfen, Shanxi Province
Shanxi Provincial Institute of Archaeology

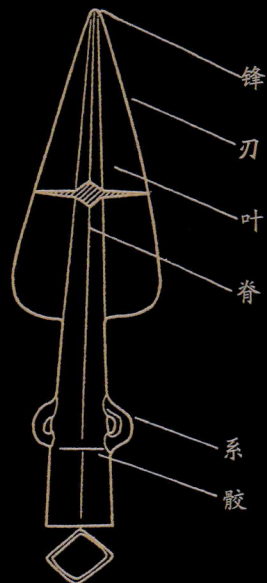

锋
刃
叶
脊
系
骹

铜 矛

Spear head

西周（公元前 1046 年－公元前 771 年）

长 21.2 厘米

山西省临汾市翼城县大河口墓地出土

山西省考古研究所藏

Western Zhou Dynasty (1046 B.C. –771 B.C.)
Length: 21.2 cm
Unearthed from Dahekou tomb complex in Yicheng
County, Linfen, Shanxi Province
Shanxi Provincial Institute of Archaeology

铜 戈

Ge dagger head

西周（公元前 1046 年 – 公元前 771 年）

长 24 厘米，阑高 7.8 厘米，内宽 3.2 厘米

山西省临汾市翼城县大河口墓地出土

山西省考古研究所藏

Western Zhou Dynasty (1046 B.C. –771 B.C.)
Length: 24 cm, height of the guard: 7.8 cm, width of the tang: 3.2 cm
Unearthed from Dahekou tomb complex in Yicheng County, Linfen, Shanxi
Province
Shanxi Provincial Institute of Archaeology

云雷纹三角援戈

Ge dagger head with cloud and thunder design

西周（公元前 1046 年 – 公元前 771 年）

长 21.1 厘米，阑高 10 厘米，内宽 5.2 厘米

山西省临汾市翼城县大河口墓地出土

山西省考古研究所藏

Western Zhou Dynasty (1046 B.C. –771 B.C.)
Length: 21.1 cm, height of the guard: 10 cm, width of the tang: 5.2 cm
Unearthed from Dahekou tomb complex in Yicheng County, Linfen, Shanxi
Province
Shanxi Provincial Institute of Archaeology

铜 戈

Ge dagger head

西周（公元前 1046 年 – 公元前 771 年）

长 21.8 厘米，阑高 4.8 厘米，内宽 3.2 厘米

山西省临汾市翼城县大河口墓地出土

山西省考古研究所藏

Western Zhou Dynasty (1046 B.C. –771 B.C.)
Length: 21.8 cm, height of the guard: 4.8 cm, width of the tang: 3.2 cm
Unearthed from Dahekou tomb complex in Yicheng County, Linfen, Shanxi
Province
Shanxi Provincial Institute of Archaeology

铜 戈

Ge dagger head

西周（公元前 1046 年 – 公元前 771 年）
长 20 厘米，阑高 10 厘米，内宽 3.1 厘米
山西省临汾市翼城县大河口墓地出土
山西省考古研究所藏

Western Zhou Dynasty (1046 B.C. –771 B.C.)
Length: 20 cm, height of the guard: 10 cm, width of the tang: 3.1 cm
Unearthed from Dahekou tomb complex in Yicheng County, Linfen, Shanxi
Province
Shanxi Provincial Institute of Archaeology

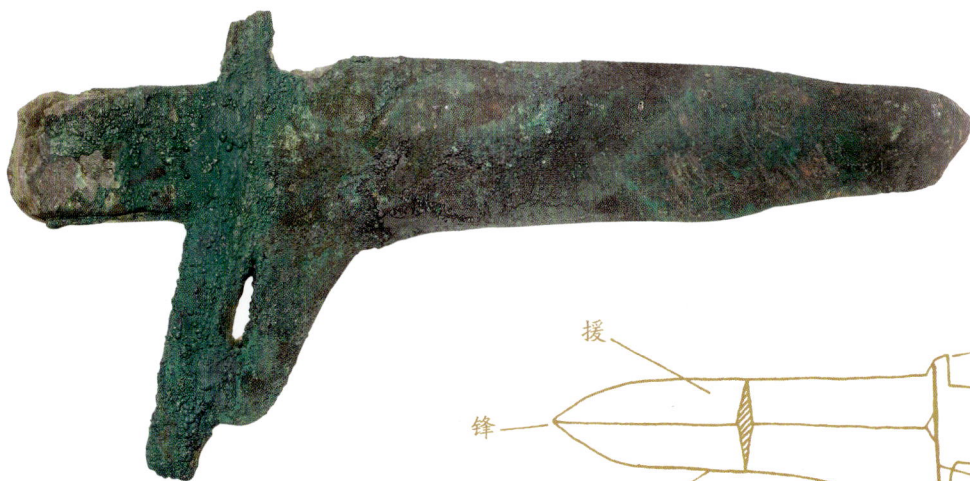

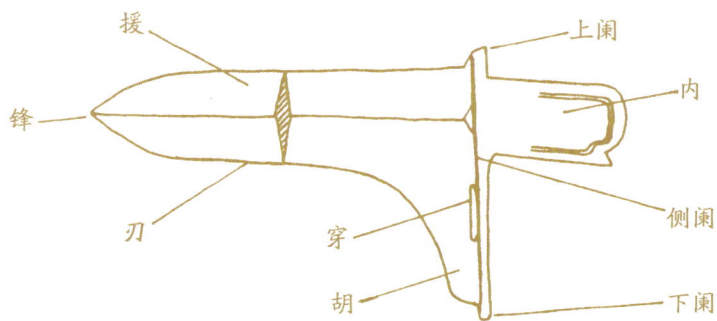

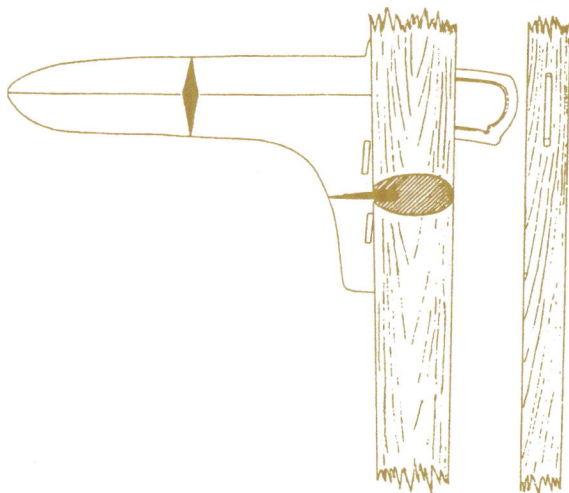

铜 戈

Ge dagger head

西周（公元前 1046 年 – 公元前 771 年）
长 19.4 厘米，阑高 6 厘米，内宽 2.8 厘米
山西省临汾市翼城县大河口墓地出土
山西省考古研究所藏

Western Zhou Dynasty (1046 B.C. –771 B.C.)
Length: 19.4 cm, height of the guard: 6 cm, width of the tang: 2.8 cm
Unearthed from Dahekou tomb complex in Yicheng County, Linfen, Shanxi
Province
Shanxi Provincial Institute of Archaeology

铜 戈

Ge dagger head

西周（公元前 1046 年 – 公元前 771 年）
长 17.7 厘米，阑高 10 厘米，内宽 3 厘米
山西省临汾市翼城县大河口墓地出土
山西省考古研究所藏

Western Zhou Dynasty (1046 B.C. –771 B.C.)
Length: 17.7 cm, height of the guard: 10 cm, width of the tang: 3 cm
Unearthed from Dahekou tomb complex in Yicheng County, Linfen, Shanxi
Province
Shanxi Provincial Institute of Archaeology

铜三角援戈

Ge dagger head with triangle blade

西周（公元前 1046 年 – 公元前 771 年）

长 23.8 厘米，阑高 13 厘米，内宽 6.5 厘米

山西省临汾市翼城县大河口墓地出土

山西省考古研究所藏

Western Zhou Dynasty (1046 B.C. –771 B.C.)
Length: 23.8 cm, height of the guard: 13 cm, width of the tang: 6.5 cm
Unearthed from Dahekou tomb complex in Yicheng County, Linfen, Shanxi
Province
Shanxi Provincial Institute of Archaeology

铜 戈

Ge dagger head

西周（公元前 1046 年 – 公元前 771 年）

长 22.3 厘米，阑高 9.5 厘米，内宽 2.7 厘米

山西省临汾市翼城县大河口墓地出土

山西省考古研究所藏

Western Zhou Dynasty (1046 B.C. –771 B.C.)
Length: 22.3 cm, height of the guard: 9.5 cm, width of the tang: 2.7 cm
Unearthed from Dahekou tomb complex in Yicheng County, Linfen,
Shanxi Province
Shanxi Provincial Institute of Archaeology

铜直内戟

Ji halberd head

西周（公元前 1046 年 – 公元前 771 年）
戈长 22.4 厘米，阑和扁形刺高 33.5 厘米，
内宽 4.3 厘米
山西省临汾市翼城县大河口墓地出土
山西省考古研究所藏

Western Zhou Dynasty (1046 B.C. −771 B.C.)
Length: 22.4 cm, height of the guard and prong: 33.5 cm, width
of the tang: 4.3 cm
Unearthed from Dahekou tomb complex in Yicheng County,
Linfen, Shanxi Province
Shanxi Provincial Institute of Archaeology

铜双龙钺

Yue battle-axe head

西周（公元前 1046 年 – 公元前 771 年）

长 14.8 厘米，刃宽 9.6 厘米，内宽 4.4 厘米

山西省临汾市翼城县大河口墓地出土

山西省考古研究所藏

Western Zhou Dynasty (1046 B.C. –771 B.C.)
Length: 14.8 cm, width of blade: 9.6 cm, width of the
tang: 4.4 cm
Unearthed from Dahekou tomb complex in Yicheng
County, Linfen, Shanxi Province
Shanxi Provincial Institute of Archaeology

铜双龙钺

Yue battle-axe head

西周（公元前 1046 年 – 公元前 771 年）

长 15 厘米，刃宽 9.7 厘米，内宽 4.4 厘米

山西省临汾市翼城县大河口墓地出土

山西省考古研究所藏

Western Zhou Dynasty (1046 B.C. –771 B.C.)
Length: 15 cm, width of blade: 9.7 cm, width of
the tang: 4.4 cm
Unearthed from Dahekou tomb complex in
Yicheng County, Linfen, Shanxi Province
Shanxi Provincial Institute of Archaeology

丧葬 事死如生

153

The Banquet Ceremony: Social Order Reflected in the Ritual Behaviors

宴饮

明君臣长幼相尊之义

鹿鳴之什

毛詩小雅

鹿鳴燕羣臣嘉賓也既飲食
賓幣帛筐篚以將其厚意然
臣嘉賓得盡其心矣呦呦鹿
野之苹我有嘉賓鼓瑟吹笙
鼓簧承筐是將人之好我示
行呦呦鹿鳴食野之蒿我有
德音孔昭視民不恌君子是
傚我有旨酒嘉賓式燕以敖
鹿鳴食野之芩我有嘉賓鼓
琴鼓瑟鼓琴和樂且湛我有
以燕樂嘉賓之心

鹿鳴

宴饮是霸国人在闲暇时举行的礼仪活动。宴礼有严格的礼仪规范，参加人员都正襟危坐，遵守"礼荐而不食，爵盈而不饮"的礼节，整个宴席只是一个虚设。这种宴饮仪式主要为了体现尊贤和养老，尊贤是治国之本，养老为安邦之本。宴饮过程安乐而有秩序，宾客尊卑分明，礼数高低有别，宴饮之人快乐而不放肆，无论长幼都得到惠泽，没有人被遗忘。霸伯做到了正身安国，才有霸国君臣的和谐与国家的长治久安。

Banquet was held as a ceremonial activity by the Ba people during their leisure time. Proper dining etiquette at a banquet was to require the attendees to behave seriously in public. In fact, the so-called banquet was just like a theater stage on which everybody was able to perform their roles in respecting the virtuous and worthy men as well as the elderly people. The banquet was only in form. The entire process of the ceremony went harmoniously and smoothly. Every single person who attended the banquet could feel happy and warm. They treated each other politely and tenderly. The concept here is that a social individual must know how to treat other social members with respect in a society, and this is like a guiding moral principle of state governance. The Count of Ba made himself an example to others in keeping the moral principles for public behavior in his state, which could guarantee that people would live a harmonious and peaceful life in the State of Ba.

玉 鹿

Jade deer

西周（公元前 1046 年 – 公元前 771 年）
高 4.7 厘米，宽 4.5 厘米，厚 0.4 厘米
山西省临汾市翼城县大河口墓地出土
山西省考古研究所藏

Western Zhou Dynasty (1046 B.C. –771 B.C.)
Height: 4.7 cm, width: 4.5 cm, thickness: 0.4 cm
Unearthed from Dahekou tomb complex in Yicheng
County, Linfen, Shanxi Province
Shanxi Provincial Institute of Archaeology

　　这两件玉鹿形态几乎完全一样，可见当时
人们对于鹿形象地认识以及玉鹿的制作技术非
常统一。

玉 鹿

Jade deer

西周（公元前 1046 年 – 公元前 771 年）
高 5.5 厘米，宽 4.6 厘米
山西省曲沃县北赵村晋侯墓地 9 号墓出土
山西博物院藏

Western Zhou Dynasty (1046 B.C. –771 B.C.)
Height: 5.5 cm, width: 4.6 cm
Unearthed from Mausoleum of Marquise of Jin,
tomb No. 9 in Beizhao Village, Quwo Courty, Shanxi
Province
Shanxi Museum

《乐记》中说"大乐与天地同和，大礼与天地同节"，在霸国的礼仪文化体系中，礼乐结合就是天地万物秩序的体现。"乐在宗庙之中，君臣上下同听之，则莫不合敬；在族长乡里之中，长幼同听之，则莫不和顺；在闺门之内中，父子兄弟同听之，则莫不和亲"，所以《孝经》中说："移风易俗，莫善于乐"，这样就做到了"四海之内合敬同爱"。

　　"高山流水"的典故向来是知音的典范，但这不是礼仪赞许的最高境界。君子聆听乐章，能从乐声中赋予新的理解。例如钟声铿锵，壮气充满，君子会想起慷慨以当的武臣；磬声清响，节义分明，君子会想起死于封疆的大臣；琴瑟之声哀怨，婉妙不越，君子会想起志义自立的大臣；竽、瑟、箫、管之声丛聚，会集揽拢，君子会想起善于蓄聚其众的大臣；鼓乐之声喧嚣，欢杂涌动，君子会想起击鼓进众的将帅之臣。这是霸国人宴饮用乐的精神追求。

山东沂南北寨出土击磬、撞钟图（局部）

甬 钟

Yongzhong chime bell

西周（公元前 1046 年 – 公元前 771 年）

通高 14 厘米，铣间距 10.7 厘米，于间宽 6.8 厘米；

甬高 4 厘米，直径 2.8 厘米

山西省临汾市翼城县大河口墓地出土

山西省考古研究所藏

Western Zhou Dynasty (1046 B.C. –771 B.C.)
Full height: 14 cm, largest width: 10.7 cm, width of the mouth:
6.8 cm; height of crown: 4 cm, crown diameter: 2.8 cm
Unearthed from Dahekou tomb complex in Yicheng County,
Linfen, Shanxi Province
Shanxi Provincial Institute of Archaeology

甬 钟

Yongzhong chime bell

西周（公元前 1046 年 – 公元前 771 年）

通高 16.5 厘米，铣间距 11.5 厘米，于间宽 7.5 厘米；甬高 5.2
厘米，直径 3.1 厘米

山西省临汾市翼城县大河口墓地出土

山西省考古研究所藏

Western Zhou Dynasty (1046 B.C. –771 B.C.)
Full height: 16.5 cm, largest width: 11.5 cm, width of the mouth:
7.5 cm; height of crown: 5.2 cm, crown diameter: 3.1 cm
Unearthed from Dahekou tomb complex in Yicheng County,
Linfen, Shanxi Province
Shanxi Provincial Institute of Archaeology

甬 钟

Yongzhong chime bell

西周（公元前 1046 年 – 公元前 771 年）
通高 18.3 厘米，铣间距 14 厘米，于间宽
9 厘米；甬高 5.5 厘米，直径 4.1 厘米
山西省临汾市翼城县大河口墓地出土
山西省考古研究所藏

Western Zhou Dynasty (1046 B.C. –771 B.C.)
Full height: 18.3 cm, largest width: 14 cm, width
of the mouth: 9 cm; height of crown: 5.5 cm,
crown diameter: 4.1 cm
Unearthed from Dahekou tomb complex in
Yicheng County, Linfen, Shanxi Province
Shanxi Provincial Institute of Archaeology

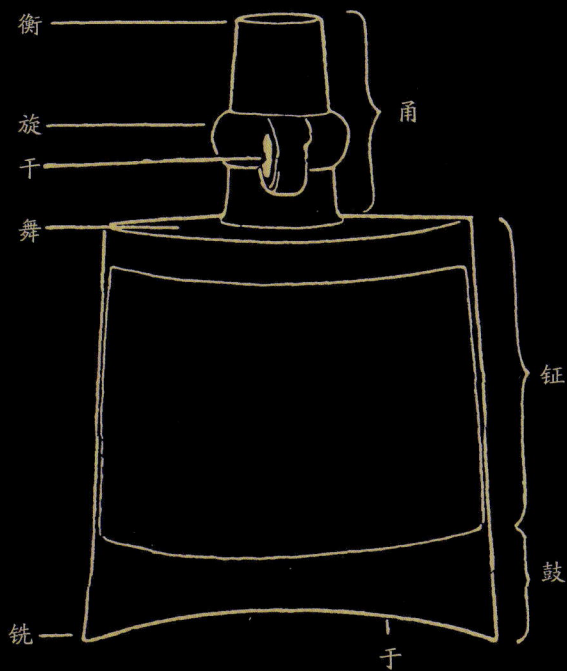

铜兽面纹铙

Nao cymbal with beast design

西周（公元前 1046 年 – 公元前 771 年）

高 15.5 厘米，铣间距 12.2 厘米，于间宽 8.6 厘米；甬高 5 厘米，直径 3.5 厘米

山西省临汾市翼城县大河口墓地出土

山西省考古研究所藏

Western Zhou Dynasty (1046 B.C. –771 B.C.)
Height: 15.5 cm, largest width: 12.2 cm, width of the mouth: 8.6 cm; height of shank: 5 cm, shank diameter: 3.5 cm
Unearthed from Dahekou tomb complex in Yicheng County, Linfen, Shanxi Province
Shanxi Provincial Institute of Archaeology

铜兽面纹铙

Nao cymbal with beast design

西周（公元前 1046 年 – 公元前 771 年）

通高 18.8 厘米，铣间距 14.7 厘米，于间宽 11 厘米；甬高 5.5 厘米，直径 3.5 厘米

山西省临汾市翼城县大河口墓地出土

山西省考古研究所藏

Western Zhou Dynasty (1046 B.C. –771 B.C.)
Height: 18.8 cm,largest width: 14.7 cm, width of the mouth: 11 cm; height of shank: 5.5 cm, shank diameter: 3.5 cm
Unearthed from Dahekou tomb complex in Yicheng County, Linfen, Shanxi Province
Shanxi Provincial Institute of Archaeology

铜兽面纹铙

Nao cymbal with beast face design

西周（公元前 1046 年 – 公元前 771 年）

通高 20.3 厘米，铙间距 17 厘米，于宽 12.2 厘米；

甬高 6.2 厘米，直径 5.6 厘米

山西省临汾市翼城县大河口墓地出土

山西省考古研究所藏

Western Zhou Dynasty (1046 B.C. –771 B.C.)
Full height: 20.3 cm, largest width: 17 cm, width of the mouth: 12.2 cm;
height of shank: 6.2 cm, shank diameter: 5.6 cm
Unearthed from Dahekou tomb complex in Yicheng County, Linfen, Shanxi Province
Shanxi Provincial Institute of Archaeology

宴饮
明君臣长幼
相尊之义

163

铜钩鑃

Goudiao chime bell

西周（公元前 1046 年 – 公元前 771 年）

通高 24 厘米，铣间距 11.2 厘米，鼓间宽 8.3 厘米，执柄长 9 厘米

山西省临汾市翼城县大河口墓地出土

山西省考古研究所藏

Western Zhou Dynasty (1046 B.C. –771 B.C.)
Full height: 24 cm, largest width: 11.2 cm, width of the mouth: 8.3 cm; height of shank: 9 cm
Unearthed from Dahekou tomb complex in Yicheng County, Linfen, Shanxi Province
Shanxi Provincial Institute of Archaeology

铜钩鑃

Goudiao chime bell

西周（公元前 1046 年 – 公元前 771 年）

通高 22 厘米，铣间距 9 厘米，鼓间宽 7 厘米，执柄长 9.5 厘米

山西省临汾市翼城县大河口墓地出土

山西省考古研究所藏

Western Zhou Dynasty (1046 B.C. –771 B.C.)
Full height: 22 cm, largest width: 9 cm, width of the mouth: 7 cm, height of shank: 9.5 cm
Unearthed from Dahekou tomb complex in Yicheng County, Linfen, Shanxi Province
Shanxi Provincial Institute of Archaeology

霸伯制定出完整而森严的礼制，以维护其统治秩序。在礼仪活动中组合使用并赋予青铜器特殊的意义，也代表使用者的身份、等级和权力。

青铜器是权力与等级的象征，其中又以鼎为最。鼎与簋、编钟及笾豆构成了霸国的礼器系统，礼器是显示人们礼仪规范最重要的器物，"贵贱有等""上下有则"是霸国等级制度的反映。

霸伯墓葬中共发现了24件青铜鼎和9件青铜簋，其中有方形的鼎2件——1件是方角方鼎，1件是圆角方鼎。随葬方鼎的墓葬都是级别较高的贵族墓葬，一般的贵族是不能随葬方鼎的。文献记载天子用九鼎八簋，诸侯用七鼎六簋，大夫用五鼎四簋等，霸伯墓葬出土的鼎簋数量为什么与文献记载不相符呢？根据考古发现表明西周中期以前尚未形成严格意义上的鼎簋制度，中期以后才形成如文献记载所说的用鼎制度。

霸伯墓葬出土的青铜器

周朝从低级的士到周天子分为五级，即低级的士、高级的士、卿大夫、诸侯、周天子，古人崇尚奇数，所以，五级分别对应的鼎的数量为一、三、五、七、九。

周天子享用九鼎，鼎内所盛之物分别为牛、羊、乳猪、鱼、干肉、牲肚、猪肉、鲜鱼、鲜干肉。

诸侯享用七鼎：比周天子少了鲜鱼和鲜干肉两项，其余七项均可享用。

卿大夫享用五鼎，即：羊、乳猪、鱼、干肉、牲肚。

高级的士享用三鼎，即：乳猪、鱼、干肉。

低级的士只能享用一鼎了，干肉。可见，干肉是当时最普遍的肉类食物了，不过就是这样的食物，也不允许一般百姓吃。所以会有"肉食者谋"这样的说法，低级的士以上的阶层才算是"肉食者"，才能有资格为国家出谋划策。

天 子 与 贵 族 的 列 鼎 数 量 及 肉 食 种 类

Specified Numbers of Ding Possessed by the Monarch and the Aristocrats in Western Zhou

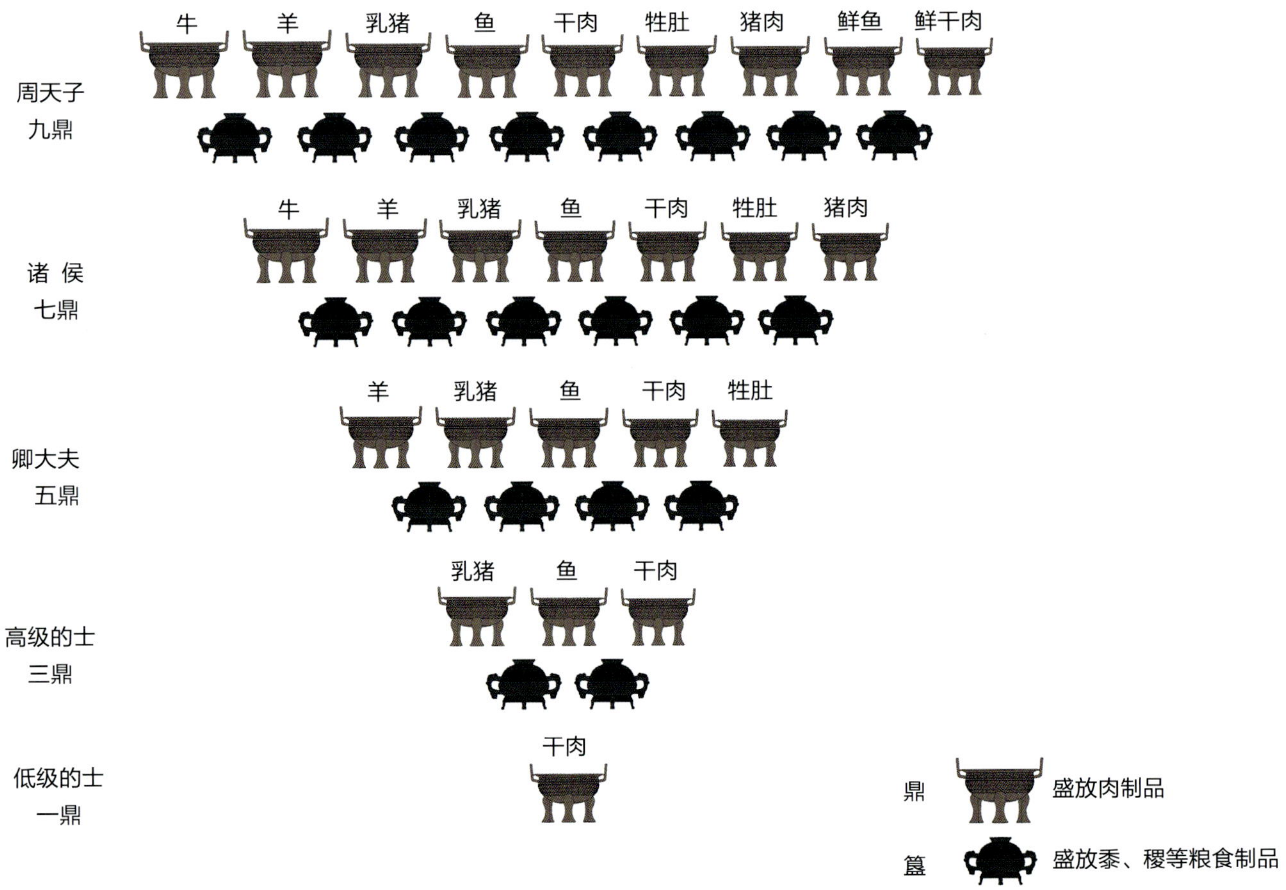

| 周天子 九鼎 | 牛 | 羊 | 乳猪 | 鱼 | 干肉 | 牲肚 | 猪肉 | 鲜鱼 | 鲜干肉 |

| 诸 侯 七鼎 | 牛 | 羊 | 乳猪 | 鱼 | 干肉 | 牲肚 | 猪肉 |

| 卿大夫 五鼎 | 羊 | 乳猪 | 鱼 | 干肉 | 牲肚 |

| 高级的士 三鼎 | 乳猪 | 鱼 | 干肉 |

| 低级的士 一鼎 | 干肉 |

鼎 盛放肉制品

簋 盛放黍、稷等粮食制品

夔凤形足温鼎

Ding tripod cauldron with *kui*-monster–phoenix shaped legs

西周（公元前 1046 年 – 公元前 771 年）

高 14.2 厘米，耳间距 11.5 厘米

山西省临汾市翼城县大河口墓地出土

山西省考古研究所藏

Western Zhou Dynasty (1046 B.C. –771 B.C.)
Height: 14.2 cm, distance between ears: 11.5 cm
Unearthed from Dahekou tomb complex in Yicheng County, Linfen, Shanxi
Province
Shanxi Provincial Institute of Archaeology

热饭的器具。

兽面纹圆鼎

Ding with beast face design

西周（公元前 1046 年 – 公元前 771 年）

高 41.8 厘米，耳间距 33.8 厘米

山西省临汾市翼城县大河口墓地出土

山西省考古研究所藏

Western Zhou Dynasty (1046 B.C. –771 B.C.)
Height: 41.8 cm, distance between ears: 33.8 cm
Unearthed from Dahekou tomb complex in Yicheng County, Linfen,
Shanxi Province
Shanxi Provincial Institute of Archaeology

器壁有铭文"伯作宝尊彝"。

兽面纹圆鼎

Ding with beast face design

西周（公元前 1046 年 – 公元前 771 年）

高 24.7 厘米，耳间距 19.4 厘米，口径 20.4 厘米

山西省临汾市翼城县大河口墓地出土

山西省考古研究所藏

Western Zhou Dynasty (1046 B.C. –771 B.C.)
Height: 24.7 cm, distance between ears: 19.4 cm, mouth
diameter: 20.4 cm
Unearthed from Dahekou tomb complex in Yicheng County,
Linfen, Shanxi Province
Shanxi Provincial Institute of Archaeology

柱足圆鼎

Ding cauldron with straight legs

西周（公元前 1046 年 – 公元前 771 年）

高 24 厘米，耳间距 20.7 厘米

山西省临汾市翼城县大河口墓地出土

山西省考古研究所藏

Western Zhou Dynasty (1046 B.C. –771 B.C.)
Height: 24 cm, distance between ears: 20.7 cm
Unearthed from Dahekou tomb complex in Yicheng
County, Linfen, Shanxi Province
Shanxi Provincial Institute of Archaeology

宴饮

相尊之义 明君臣长幼

四足带盖方鼎

Square *ding* cauldron with four legs

西周（公元前 1046 年－公元前 771 年）

高 28 厘米，耳间距 22.7 厘米

山西省临汾市翼城县大河口墓地出土

山西省考古研究所藏

Western Zhou Dynasty (1046 B.C. –771 B.C.)
Height: 28 cm, distance between ears: 22.7 cm
Unearthed from Dahekou tomb complex in Yicheng
County, Linfen, Shanxi Province
Shanxi Provincial Institute of Archaeology

兽面云雷纹鼎

Ding tripod cauldron with cloud, thunder and beast face design

西周（公元前 1046 年－公元前 771 年）

高 26.5 厘米，耳间距 20.8 厘米

山西省临汾市翼城县大河口墓地出土

山西省考古研究所藏

Western Zhou Dynasty (1046 B.C. – 771 B.C.)
Height: 26.5 cm, distance between ears: 20.8 cm
Unearthed from Dahekou tomb complex in Yicheng
County, Linfen, Shanxi Province
Shanxi Provincial Institute of Archaeology

兽面纹圆鼎

Ding tripod cauldron with beast face design

西周（公元前 1046 年 – 公元前 771 年）

高 20.5 厘米，耳间距 16 厘米

山西省临汾市翼城县大河口墓地出土

山西省考古研究所藏

Western Zhou Dynasty (1046 B.C. –771 B.C.)
Height: 20.5 cm, distance between ears: 16 cm
Unearthed from Dahekou tomb complex in Yicheng County,
Linfen, Shanxi Province
Shanxi Provincial Institute of Archaeology

龙纹高圈足簋

Gui food container with dragon design and circular base

西周（公元前 1046 年 – 公元前 771 年）

高 14.3 厘米，耳间距 25.2 厘米，圈足径 14.4 厘米

山西省临汾市翼城县大河口墓地出土

山西省考古研究所藏

Western Zhou Dynasty (1046 B.C. –771 B.C.)
Height: 14.3 cm, distance between ears: 25.2 cm, base diameter: 14.4 cm
Unearthed from Dahekou tomb complex in Yicheng County, Linfen, Shanxi Province
Shanxi Provincial Institute of Archaeology

宴饮

明君臣长幼

相尊之义

龙纹高圈足簋

Gui food container with dragon design and circular base

西周（公元前 1046 年 – 公元前 771 年）

高 14.3 厘米，耳间距 25.2 厘米，圈足径 14.2 厘米

山西省临汾市翼城县大河口墓地出土

山西省考古研究所藏

Western Zhou Dynasty (1046 B.C. –771 B.C.)
Height: 14.3 cm, distance between ears: 25.2 cm, base diameter: 14.2 cm
Unearthed from Dahekou tomb complex in Yicheng County, Linfen, Shanxi Province
Shanxi Provincial Institute of Archaeology

高圈足簋

Gui food container with circular base

西周（公元前 1046 年 – 公元前 771 年）

高 14.5 厘米，耳间距 26.5 厘米，圈足径 14.2 厘米

山西省临汾市翼城县大河口墓地出土

山西省考古研究所藏

Western Zhou Dynasty (1046 B.C. –771 B.C.)
Height: 14.5 cm, distance between ears: 26.5 cm, base diameter: 14.2 cm
Unearthed from Dahekou tomb complex in Yicheng County, Linfen, Shanxi Province
Shanxi Provincial Institute of Archaeology

方座铃簋

Gui food container with square base

西周（公元前 1046 年 – 公元前 771 年）
高 21.3 厘米，耳间距 26.8 厘米，底座
17 厘米 ×15.8 厘米
山西省临汾市翼城县大河口墓地出土
山西省考古研究所藏

Western Zhou Dynasty (1046 B.C. – 771 B.C.)
Height: 21.3 cm, distance between ears: 26.8 cm,
base: 17 cm×15.8 cm
Unearthed from Dahekou tomb complex in
Yicheng County, Linfen, Shanxi Province
Shanxi Provincial Institute of Archaeology

174

在宴饮时，先要将牛、羊、猪等在镬（最大的锅）中煮熟，然后用匕（头部尖锐的取食器）取出来，放入鼎内，调和入味。为了保温和防灰，要加上盖子。将鼎从庖厨移送到行礼的场所，是用"铉"贯穿鼎的两耳抬走，鼎不是食器，所以食用之前，要再次用匕将肉从鼎中取出，放在俎上，然后再陈设在食案上。鼎与俎是配套使用的，所以在礼器的组合中，两者的数量总是相同。

盛食器还有簋、笾、豆。簋是盛稻粱的圆形器皿，也有极少量是方形或长方形的，有盖。在礼器的组合中，鼎与簋最为重要，但前者用奇数，后者用偶数。笾与豆的形状相似。笾是盛肉干、果实等干燥的食物用的，豆则是盛腌渍的蔬菜、肉酱等有汁的食物用的。笾与豆通常配合使用，而且都用双数，所以《礼记》说"鼎俎奇而笾豆偶"。

兽面纹甗

Yan steamer with beast face design

西周（公元前 1046 年 – 公元前 771 年）
高 35 厘米，耳间距 22.3 厘米
山西省临汾市翼城县大河口墓地出土
山西省考古研究所藏

Western Zhou Dynasty (1046 B.C. –771 B.C.)
Height: 35 cm, distance between ears: 22.3 cm
Unearthed from Dahekou tomb complex in Yicheng
County, Linfen, Shanxi Province
Shanxi Provincial Institute of Archaeology

甗，音演。炊蒸器，中部有箅（音必）子。

铜盨

Xu food container

西周（公元前 1046 年 - 公元前 771 年）

长 28 厘米，高 17.5 厘米，宽 17 厘米

山西省临汾市翼城县大河口墓地出土

山西省考古研究所藏

Western Zhou Dynasty (1046 B.C. –771 B.C.)
Length: 28 cm, height: 17.5 cm, width: 17 cm
Unearthed from Dahekou tomb complex in Yicheng
County, Linfen, Shanxi Province
Shanxi Provincial Institute of Archaeology

分裆鬲
Li steamer

西周（公元前 1046 年 – 公元前 771 年）
高 17.2 厘米，耳间距 15 厘米
山西省临汾市翼城县大河口墓地出土
山西省考古研究所藏

Western Zhou Dynasty (1046 B.C. –771 B.C.)
Height: 17.2 cm, distance between ears: 15 cm
Unearthed from Dahekou tomb complex in
Yicheng County, Linfen, Shanxi Province
Shanxi Provincial Institute of Archaeology

煮饭用器。

弦纹鬲
Li steamer with string pattern

西周（公元前 1046 年 – 公元前 771 年）
高 14 厘米，耳间距 11 厘米
山西省临汾市翼城县大河口墓地出土
山西省考古研究所藏

Western Zhou Dynasty (1046 B.C. –771 B.C.)
Height: 14 cm, distance between ears: 11 cm
Unearthed from Dahekou tomb complex in Yicheng County,
Linfen, Shanxi Province
Shanxi Provincial Institute of Archaeology

铜鬲

Li steamer

西周（公元前 1046 年 – 公元前 771 年）

高 17 厘米，耳间距 14.5 厘米

山西省临汾市翼城县大河口墓地出土

山西省考古研究所藏

Western Zhou Dynasty (1046 B.C. –771 B.C.)
Height: 17 cm, distance between ears: 14.5 cm
Unearthed from Dahekou tomb complex in Yicheng
County, Linfen, Shanxi Province
Shanxi Provincial Institute of Archaeology

铜鬲

Li steamer

西周（公元前 1046 年 – 公元前 771 年）

高 18 厘米，耳间距 14 厘米

山西省临汾市翼城县大河口墓地出土

山西省考古研究所藏

Western Zhou Dynasty (1046 B.C. –771 B.C.)
Height: 18 cm, distance between ears: 14 cm
Unearthed from Dahekou tomb complex in Yicheng
County, Linfen, Shanxi Province
Shanxi Provincial Institute of Archaeology

　　"在朝序爵，在乡序齿"，朝廷中以官爵大小为序，而民间不然，是以年龄为序，少不越长。宴饮中有尊老、敬老的规定。六十岁以上的老人方可坐着饮酒。在饮食上对老人也有所优礼，五十岁的人可以吃细粮，六十岁的人有预备的肉食，七十岁的人每餐应该有两个好菜，八十岁的人应该常吃美食，九十岁的人饮食都在寝室。餐具的使用也是按照年龄配设不等的豆数：六十岁者三豆，七十岁者四豆，八十岁者五豆，九十岁者六豆。豆内所盛，是奉养老人的食物。豆数不同，则所受到的奉养也不同。

　　除了家庭的照顾之外，老人还可得到国家的关心，霸国设立养老机构让老人颐养天年。在宴饮中懂得了尊长养老的道理，回家就会有孝悌的行动。在家里懂得孝悌，出外懂得尊长养老，就能形成良好的风教。有了良好的社会风教，国家就安定了。

　　礼器中的酒器，又可以分为盛酒器和饮酒器两大类。盛酒器主要有尊、罍、卣、壶、缶等。霸伯用尊盛酒，卿大夫用方壶，未取得爵位的士用圆壶。饮酒器有爵、觯（音质）、觚（音孤）、觥（音宫）等。爵容为一升酒，觚为二升，觯为三升。觥在饮酒器中容量最大，所以在君臣宴饮等场合，常常用作罚酒之器。

椁室内上层青铜器

“内父丁”爵

Jue goblet with inscription “Nei Fu Ding”

西周（公元前 1046 年 – 公元前 771 年）

高 21 厘米，柱间距 7.5 厘米，尾流长 19 厘米

山西省临汾市翼城县大河口墓地出土

山西省考古研究所藏

Western Zhou Dynasty (1046 B.C. – 771 B.C.)
Height: 21 cm, distance between legs: 7.5 cm, length of spout: 19 cm
Unearthed from Dahekou tomb complex in Yicheng County, Linfen, Shanxi Province
Shanxi Provincial Institute of Archaeology

这件爵鋬后器腹外壁有铭文“内父丁”，为温酒器或饮酒器，夏商两周时期较盛行，爵位之爵即由此而来。

父乙爵

Jue goblet with inscription “Fu Yi”

西周（公元前 1046 年 – 公元前 771 年）

高 19.5 厘米，柱间距 7 厘米，尾流长 16.5 厘米

山西省临汾市翼城县大河口墓地出土

山西省考古研究所藏

Western Zhou Dynasty (1046 B.C. – 771 B.C.)
Height: 19.5 cm, distance between legs: 7 cm, length of spout: 16.5 cm
Unearthed from Dahekou tomb complex in Yicheng County, Linfen, Shanxi Province
Shanxi Provincial Institute of Archaeology

兽形鋬后器外壁有铭文“父乙”二字，一柱有铭文“[illegible]作”。

云雷纹爵

Jue goblet with cloud and thunder pattern

西周（公元前 1046 年 – 公元前 771 年）

高 22.2 厘米，柱间距 9.1 厘米，尾流长 17 厘米

山西省临汾市翼城县大河口墓地出土

山西省考古研究所藏

Western Zhou Dynasty (1046 B.C. –771 B.C.)
Height: 22.2 cm, distance between legs: 9.1 cm, length of spout: 17 cm
Unearthed from Dahekou tomb complex in Yicheng County, Linfen, Shanxi Province
Shanxi Provincial Institute of Archaeology

龙纹爵

Jue goblet with dragon design

西周（公元前 1046 年 – 公元前 771 年）

高 21.5 厘米，柱间距 9 厘米，尾流长 16 厘米

山西省临汾市翼城县大河口墓地出土

山西省考古研究所藏

Western Zhou Dynasty (1046 B.C. –771 B.C.)
Height: 21.5 cm, distance between legs: 9 cm, length of spout: 16 cm
Unearthed from Dahekou tomb complex in Yicheng County, Linfen, Shanxi Province
Shanxi Provincial Institute of Archaeology

温酒器或饮酒器。

兽面纹爵

Jue goblet with beast face design

西周（公元前 1046 年 – 公元前 771 年）

高 22.5 厘米，柱间距 9 厘米，尾流长 18 厘米

山西省临汾市翼城县大河口墓地出土

山西省考古研究所藏

Western Zhou Dynasty (1046 B.C. –771 B.C.)
Height: 22.5 cm, distance between legs: 9 cm, length of spout: 18 cm
Unearthed from Dahekou tomb complex in Yicheng County, Linfen, Shanxi Province
Shanxi Provincial Institute of Archaeology

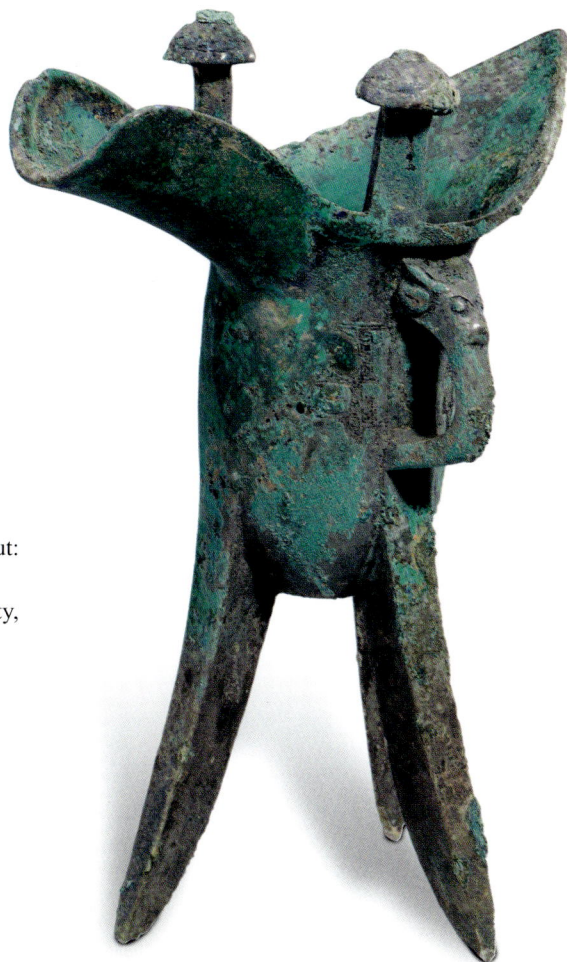

"析父丁"觯

Zhi wine vessel with inscription "Xi Fu Ding"

西周（公元前 1046 年 – 公元前 771 年）

高 21 厘米，口径 9.7 厘米 ×8.8 厘米，底径 9 厘米 ×7.7 厘米

山西省临汾市翼城县大河口墓地出土

山西省考古研究所藏

Western Zhou Dynasty (1046 B.C. – 771 B.C.)
Height: 21 cm, mouth diameter: 9.7 cm×8.8 cm, bottom diameter: 9 cm×7.7 cm
Unearthed from Dahekou tomb complex in Yicheng County, Linfen, Shanxi Province
Shanxi Provincial Institute of Archaeology

铜 觯

Zhi wine vessel

西周（公元前 1046 年－公元前 771 年）

高 14.5 厘米，口径 5.5 厘米，底径 5.5 厘米

山西省临汾市翼城县大河口墓地出土

山西省考古研究所藏

Western Zhou Dynasty (1046 B.C. – 771 B.C.)
Height: 14.5 cm, mouth diameter: 5.5 cm, bottom diameter:
5.5 cm
Unearthed from Dahekou tomb complex in Yicheng County,
Linfen, Shanxi Province
Shanxi Provincial Institute of Archaeology

弦纹觯

Zhi wine vessel with string pattern

西周（公元前 1046 年－公元前 771 年）

高 17.2 厘米，口径 8.7 厘米，底径 6.1 厘米

山西省临汾市翼城县大河口墓地出土

山西省考古研究所藏

Western Zhou Dynasty (1046 B.C. – 771 B.C.)
Height: 17.2 cm, mouth diameter: 8.7 cm, bottom diameter: 6.1 cm
Unearthed from Dahekou tomb complex in Yicheng County,
Linfen, Shanxi Province
Shanxi Provincial Institute of Archaeology

兽头凤鸟纹尊

Zun wine vessel with beast head and phoenix design

西周（公元前 1046 年 – 公元前 771 年）

高 18 厘米，口径 17.4 厘米

山西省临汾市翼城县大河口墓地出土

山西省考古研究所藏

Western Zhou Dynasty (1046 B.C. – 771 B.C.)
Height: 18 cm, mouth diameter: 17.4 cm
Unearthed from Dahekou tomb complex in Yicheng County, Linfen,
Shanxi Province
Shanxi Provincial Institute of Archaeology

盛酒器。

凤鸟纹提梁卣

You wine container with overtop handle and phoenix design

西周（公元前 1046 年－公元前 771 年）

高 20.5 厘米，耳间距 17.5 厘米

山西省临汾市翼城县大河口墓地出土

山西省考古研究所藏

Western Zhou Dynasty (1046 B.C. – 771 B.C.)
Height: 20.5 cm, distance between ears: 17.5 cm
Unearthed from Dahekou tomb complex in Yicheng County, Linfen, Shanxi
Province
Shanxi Provincial Institute of Archaeology

　　盛酒器。用于赏赐有功的诸侯。《诗经·大雅·江汉》中有"厘尔圭瓒，秬鬯一卣"，意为"赏你玉杓世世传，黍酒一壶香又甜"。秬鬯（音具畅），一种香酒。

铜 盘

Pan basin

西周（公元前 1046 年 – 公元前 771 年）

高 11.5 厘米，口径 34 厘米，圈足径 19 厘米

山西省临汾市翼城县大河口墓地出土

山西省考古研究所藏

Western Zhou Dynasty (1046 B.C. –771 B.C.)
Height: 11.5 cm, mouth diameter: 34 cm, circular base diameter: 19 cm
Unearthed from Dahekou tomb complex in Yicheng County, Linfen, Shanxi Province
Shanxi Provincial Institute of Archaeology

霸国贵族宴饮的礼仪较多，比如他们吃饭是分餐制；酒具在宴饮的过程中要多次洗涤；由于吃饭不用筷子，而是用手抓着来吃，因此在宴饮前后需要净手，净手时一人注水，一人承盘接水。商周时期的墓葬中，盉与盘是一套器物，有盉必有盘。盉最早是用来调酒的，后来也用来注水，有的青铜器也存在着一器多用的情况。盘用来承接水。西周晚期匜代替了盉，匜与盉的功能相同。

四足盉

He pot with four legs and a lid

西周（公元前 1046 年 – 公元前 771 年）
高 25 厘米
山西省临汾市翼城县大河口墓地出土
山西省考古研究所藏

Western Zhou Dynasty (1046 B.C. –771 B.C.)
Height: 25 cm
Unearthed from Dahekou tomb complex in Yicheng County, Linfen, Shanxi Province
Shanxi Provincial Institute of Archaeology

酒器或水器。

凤鸟纹链盖盉

He wine container with phoenix design and a lid

西周（公元前 1046 年 – 公元前 771 年）

高 25 厘米，流錾间距 28 厘米

山西省临汾市翼城县大河口墓地出土

山西省考古研究所藏

Western Zhou Dynasty (1046 B.C. –771 B.C.)
Height: 25 cm, distance between spout and handle: 28 cm
Unearthed from Dahekou tomb complex in Yicheng County, Linfen,
Shanxi Province
Shanxi Provincial Institute of Archaeology

瓦棱纹匜

Yi ewer with tile ridge pattern

西周（公元前 1046 年－公元前 771 年）

高 13.5 厘米，通长 25.6 厘米，最宽处 12.2 厘米

山西省临汾市翼城县大河口墓地出土

山西省考古研究所藏

Western Zhou Dynasty (1046 B.C. –771 B.C.)
Height: 13.5 cm, full length: 25.6 cm, largest width: 12.2 cm
Unearthed from Dahekou tomb complex in Yicheng County,
Linfen, Shanxi Province
Shanxi Provincial Institute of Archaeology

　　注水器，与盘配合使用。商周时期宴饮时，饭食前后沃盥之用。

附耳圈足盘

Pan basin with cicular base and ears

西周（公元前 1046 年 – 公元前 771 年）

高 6.5 厘米，耳间距 22.5 厘米，圈足径 13 厘米

山西省临汾市翼城县大河口墓地出土

山西省考古研究所藏

Western Zhou Dynasty (1046 B.C. –771 B.C.)
Height: 6.5 cm, distance between ears: 22.5 cm, base diameter: 13 cm
Unearthed from Dahekou tomb complex in Yicheng County, Linfen,
Shanxi Province
Shanxi Provincial Institute of Archaeology

承水器。

鸟形盉

He ewer in the shape of a bird

西周（公元前 1046 年 – 公元前 771 年）

高 35.2 厘米，流尾长 37 厘米，宽 19 厘米

山西省临汾市翼城县大河口墓地出土

山西省考古研究所藏

Western Zhou Dynasty (1046 B.C. –771 B.C.)
Height: 35.2 cm, length from spout to handle: 37 cm, width: 19 cm
Unearthed from Dahekou tomb complex in Yicheng County, Linfen,
Shanxi Province
Shanxi Provincial Institute of Archaeology

  盖内有铭文 8 行 51 字，自名为盉。鸟形盉的发现为商周时期青铜器增加了一种新器形。盉多与盘配套用于祭祀或宴饮活动中浇水洗手，这座墓里恰好出土了一件铜盘与这件鸟形盉相配，可惜盘已残。

  对于鸟形盉铭文的释读目前尚不统一，清华大学出土文献研究与保护中心的李学勤教授释读如下：乞立誓说："我所作谋议如果不合君命，而是我自己私行策划，就受鞭刑。"乞亲自乘有车蔽的传车前往各地，重复所立誓言，说："我已立誓要上合君命，假如我违反誓辞，便应该遭到流弃，使君命仍得执行。"乞因此铸造盘盉，传于子孙使用。

呦呦鹿鸣

燕国公主眼里的霸国

Harmonious Life: The State of
Ba in the Eyes of a Yan Princess

结 语

《礼记·礼运》中曰："大道之行也，天下为公，选贤与能，讲信修睦。故人不独亲其亲，不独子其子，使老有所终，壮有所用，幼有所长，鳏、寡、孤、独、废疾者皆有所养，男有分，女有归。货恶其弃于地也，不必藏于己；力恶其不出于身也，不必为己。是故谋闭而不兴，盗窃乱贼而不作，故外户而不闭，是谓大同。"

译　文

《礼记·礼运》中说："在大道施行的时候，天下是人们所共有的，把品德高尚的人、能干的人选拔出来，（人人）讲求诚信，致力和睦，所以人们不只是敬爱自己的双亲，不只是疼爱自己的子女，更能博爱世人，使老年人能安享晚年，使壮年人能为社会效力，使孩子健康成长，使老而无妻的人、老而无夫的人、幼而无父的人、老而无子的人、残疾人都有所养护。男子有职务，女子有归宿。对于财货，就担心它丢弃在地上得不到合理利用，却不一定要自己私藏；人们都愿意为公众之事竭尽全力，而不一定为自己谋取私利。因此不会有为非作歹的念头，盗窃掠夺的事情自然不会发生，所以外门无须紧闭。这就叫做大同社会。"

Epilogue

The Record of Rites goes as follows:

When the perfect order prevails, the world is like a home shared by all. Virtuous and worthy men are elected to public office, and capable men hold posts of gainful employment in society; peace and trust among all men are the maxims of living. All men love and respect their own parents and children, as well as the parents and children of others. This is caring for the old; there are jobs for adults; there are nourishment and education for the children. There is a means of support for the widows, and the widowers; for all who find themselves alone in the world; and for the disabled. Every man and woman has an appropriate role to play in the family and society. A sense of sharing displaces the effects of selfishness and materialism. A devotion to public duty leaves no room for idleness. Intrigues and conniving for ill gain are unknown. Villains such as thieves and robbers do not exist. The door to every home need never be locked and bolted day or night. These are the characteristics of an ideal world, the commonwealth state.